Canada's Water, Yours to Protect

A Primer on Planning Together

Sandra E. Smith

Sandra Smith holds a PhD in Geography (University of Victoria), an MA in Community and Regional Planning (University of British Columbia) and is a Member (Ret) of the Canadian Institute of Planners. Her water experience was with BC Environment's Water Management Program and island planning experience as Manager, Local Planning Services for the Islands Trust. Dr. Smith was an Adjunct Assistant Professor at the University of Victoria's Department of Geography and taught Water Management and Urban Social Geography and Planning.

Robert France is the Associate Professor of Watershed Management at Dalhousie University and is one of twelve members of the Water Partner Advisory Committee to the Council of the Federation of Canada. He is the author or editor of 18 books, including *Deep Immersion: The Experience of Water*, *Handbook of Water Sensitive Planning and Design*, *Facilitating Watershed Management: Fostering Awareness and Stewardship*, and *Introduction to Watershed Development: Understanding and Managing the Impacts of Sprawl*, as well as more than two hundred technical papers.

To the memory of Jim McCracken
 —*Water Manager, Mentor*

Canada's Water, Yours to Protect:
A Primer on Planning Together

≈

Sandra E. Smith

ISBN 978-1-927043-32-5

Canada's Water, Yours to Protect: A Primer on Planning Together

First Edition, April 2014

All rights reserved.

Copyright © 2014 by Sandra E. Smith and Green Frigate Books

www.greenfrigatebooks.com (Winnipeg, Manitoba)

Smith, Sandra E.
Canada's Water, Yours to Protect: A Primer on Planning Together
- Environmental Science. 2. Water Management. 3. Canada.

By using a recycled paper, the environmental impact was reduced by.... 2.62 trees (40 foot tall and 6-8" diameter), 1,111 gallons of water, 4.47 kilowatt hours of electricity, 122 pounds of solid waste and 241 lbs of greenhouse gases.

Cover Photo: Waterfall on Louise Creek, Alberta by Megan Lorenz
www.mlorenzphotography

Printed in Canada

ACKNOWLEDGEMENTS

To those who enabled and informed my water policy adventure: Peter Brady, Don Kasianchuk, Jon O'Riordan, Tony Dorcey, Harriet Rueggeberg, Livia Meret and Bill Hollingshead; and to BC Environment's water management program, each and all.

To the politicians and staff of the Islands Trust and to the people of Hornby Island who protect their groundwater.

To the University of Victoria's Geography Department and Library facilities; and to past students, with special thanks to Anna Thomas Peacock, Christine Webb and Michael Zbarsky.

To readers who substantially improved this book through review of early drafts.

To Robert France, with sincere gratitude for his inspiration and thoughtful comments.

To Green Frigate Books and to Garrett Brown and Jennifer Brown of Opaque Design and Print Production for their magic in transforming manuscript to book.

To Jane Henderson for sharing her story.

To Claire Hastings for permission to use the Connor Youngerman quotation.

To Doug Porteous for his wise counsel.

To those who never failed to encourage and assist: Judith Allen, Thomas Blair, Jill and Duart Campbell, Kim Fowler, Fran Gundry, Georgina and Irwin Henderson, Marg Palmer, Sarah Pollard, Robert O'Regan, Janet Wood and Henry Wiseman, and Carmie Verdone.

For their never-ending patience and care, to my husband Peter, son Jeremy, daughter-in-law Tara, and their Westie, Ms Lucy who, true to her breed, keeps us marching in line.

CONTENTS

Acknowledgements	vii
FOREWORD: Protecting The Health and Future of Canada's Most Important Natural Resource by Robert L. France	1
PROLOGUE: A Personal Journey	9
CHAPTER 1: Introduction	11
Canada's Water, Yours to Protect	12
How Did We Get to Here?	13
Organization of the Book	14
CHAPTER 2: Being Involved	17
What is Water Management?	17
Water: A Short Description of a Critical Resource	18
Issues and Actors/Influencers	21
Opportunities for Involvement	27
THEME 1: WATER AVAILABILITY	31
CHAPTER 3: Allocating Surface Water: Is There a Better Way?	33
Water Allocation Conflicts	34
How is Water Used?	34
Water Allocation Schemes in Canada	35
Challenges to the Allocation Schemes	44
Conclusion	46
CHAPTER 4: Protecting Groundwater: A Community's Story	47

x   Contents

Groundwater Defined 47
Groundwater Use 48
Groundwater Governance and Management 50
BC's Islands Trust Area and Groundwater Dependence 51
Protecting Groundwater on Hornby Island 51
Learning from the Hornby Island Experience 58
Future Directions and Conclusion 59

CHAPTER 5: Finding an Effective Way to Curb Water Use 61
Water's Worth 62
Water: A Special Economic Good 63
What is Water Demand Management? 64
Full Cost Pricing 65
Why Not Increase the Price of Water? 68
Water Efficiency and Conservation Measures 69
Which is Better? 71
Actions to Encourage Water Demand Management and Augment Supplies   72
Future Directions and Conclusion 74

THEME 2: PROTECTING PUBLIC SAFETY AND WATER'S
ECOLOGICAL AND INTRINSIC VALUES 77

CHAPTER 6: Who Speaks For The Fish and Who For The Stream? 79
Problems Affecting the Aquatic Ecosystem 79
Aquatic Ecosystem Protection 84
Honouring the Intangible Values of Water 91
Future Actions 94
Conclusion 95

CHAPTER 7: Protecting Drinking Water Quality 97
Problems Defined 98
Effect of Drinking Water Quality Problems 99
Government Action 101
Non-Government Activities 106
Future Actions and Conclusion 107

CHAPTER 8: Flooding: Not a Question of If, But When 109
Experiencing a Flood 110
Flooding Events 111
Floodplain Management: Reacting to the Flood Hazard 114
Minimizing Flood Damages 115

Other Considerations	120
Conclusion	121

THEME 3: WATER DEVELOPMENT AND INFRASTRUCTURE — 123

CHAPTER 9: Dams: Can We Find a Balanced View? — 125
The Pervasiveness of Dams	126
The Benefits of Dam Construction	127
The Problem with Dams	129
Other Considerations	136
Conclusion	139

CHAPTER 10: Water, Water Going Everywhere — 141
Water Conflict	141
What are the Reasons for Concern?	142
Provincial Government Involvement in Water Protection	149
The Federal Response	154
Other Considerations	158
Conclusion	161

CHAPTER 11: Planning Together — 163
Water Management Planning and Integrated Resource Management	163
Types of Plans	165
Strategic Planning for Water Organizations	166
Water Availability Planning	167
Protection Plans	168
Water Development and Infrastructure Plans	170
Multiple Issues (Basin to Watershed)	174
Future Directions	177
Conclusion	178

CHAPTER 12: Caring More — 181
What Next?	182
Conclusion	190

BIBLIOGRAPHY — 191

APPENDIX 1: GOVERNMENT'S WATER MANAGEMENT ROLES — 225

ENDNOTES — 231

LIST OF FIGURES

Figure 2-1: Water Management Framework	19
Figure 2-2: Change in Emphasis over Time	20
Figure 3-1: Riparian Doctrine	37
Figure 3-2: Prior Appropriation Scheme	38

LIST OF TABLES

Table 6-1: Possible Tools for Provincial-Level Aquatic Ecosystem Protection	86
Table 6-2: Local-Level Land Use Planning Tools to Protect the Aquatic Ecosystem	88
Table 10-1: Water Export: The Headline Writer's Art	148
Table 10-2: Media Frame the Water Export Debate	149

INDEX   247

FOREWORD

Protecting The Health and Future of Canada's Most Important Natural Resource

Canada is a country whose very history is defined by water. Not only do we have the Great Lakes—St. Lawrence system, every bit as important to developing our nationhood as the Nile, the Tigris and Euphrates, the Indus, and the Yellow and Yangtze Rivers were to Egypt, Iraq, Pakistan/India, and China, but we also have that complicated ribbon of blue ways that linked both the Lakehead and Hudson Bay to the western and northern extremes of our country. Here through the bold endeavors of that pantheon of intrepid early explorers and later entrepreneurial fur merchants, following as they did the watery trails of untold generations of First Nations Peoples, Canada was connected and was born (Innis 1962; Morse 1969; Newman 1986, 1987; Anon. 1997; Huck et al. 2002). No wonder then that for many the Canadian spirit and soul lies in sitting (and sometimes much more!) in a canoe atop our multitude of lakes and rivers (Benidickson 1997; Jennings et al. 1999; Raffan 1999), possibly while enjoying a retreat at the summer cottage (Casey 2009; Harrison 2013). In consequence, we treasure and celebrate the rich palimpsest landscape of our multi-storied waters (Unwin 2003; Robertson and McCracken 2003; Russell 2004; Evenden 2007; Armstrong et al. 2010; Mihell 2012).

But all is not well in our relationship with water (France 2003), either on a global scale (France 2013) or in Canada in particular, where some have gone as far as to posit that we are actually in a "crisis" (Brooymans 2011). Whether acute events such as dramatic floods and poisonings or more subtle chronic problems, there is no doubt that the health and future of our most precious natural resource is at threat. No surprise then that more than half a dozen research centres at Canadian universities have been established with a specific focus on water management concerns. And in 2010, the Council of the Federation (COF) created a *Water Charter* (COF 2010) based on the following high-minded truisms:

- Water is a natural resource that is an essential component of all life on earth and there is no substitute for water
- Adequate clean water is critical to human health, sanitation and the liveability of communities across Canada
- Water in its natural state is critical for supporting ecosystem health, maintaining fisheries, providing recreation and attracting tourism
- Water is an essential input to agriculture and is necessary for industry and resource development
- The changing climate is already affecting this vital resource
- Canadians recognize our collective obligation to be responsible global water stewards and the need to continue to improve our water conservation and water quality preservation and enhancement efforts
- Canadians have the potential to help address global water issues by developing and commercializing innovative new technologies and services, and to be leaders in the development and sale of new technologies and services to improve water conservation and quality across the world
- Innovation and the development of commercial potential for technology will help improve water efficiency across Canada while, leveraging our existing strengths, boosting economic opportunity, and positioning the country as a global leader in the field
- Provinces and territories recognize that many watersheds do not follow national, provincial and territorial boundaries
- Provinces and territories recognize we can improve our efforts by working in partnership and leveraging the successes in the management of water conservation and water quality protection in our individual jurisdictions
- Working to achieve overarching water goals of reducing water consumption and increasing water efficiency; protecting our water quality and adapting to effects of climate change on water can have both environmental and economic benefits and is essential to a healthy, secure, and prosperous Canada

The Water Charter is administered by a newly established *Water Stewardship Council* (COF 2011a) consisting of one Deputy Minister from each Canadian province and territory brought together to undertake all timely measures in the respective jurisdictions to:

- Reinforce water conservation, water quality and adapting to the water-related aspects of climate change as a key priority for businesses, citizens, and their governments
- Make water use more efficient, beginning by challenging water use sectors to prepare water conservation and efficiency plans
- Enhance our water monitoring effort and cooperate and share information on water conservation and water quality
- Work with municipalities to ensure they have plans to deal with water-related emergencies and enhance best-practice sharing of planning tools among communities
- Encourage Canadians and Canadian companies to be leaders in the development and sale of new technologies and services for water conservation and protection
- Work with public and private sector groups to make World Water Day a national event with a visible and coordinated focus on the above priorities
- Collaborate with Provinces, Territories and State governments on trans-boundary issues

But the governmental WSC recognized that it could not undertake these actions alone without input from key stakeholders. Because of this, the COF in 2011 created a *Water Partner Advisory Committee* (WPAC) composed of twelve stakeholder members having expertise and knowledge on water conservation, water sustainability, water technology, and/or climate change (COF 2011b). These WPAC representatives, of which I am a member, were assembled from industry, local interest groups, academia, business, First Nations, and NGOs with the goal of providing expert advice from a broad-based stakeholder perspective to the WSC on water use, water quality and climate change in an integrated and holistic manner

In 2011, members of the WPAC met to identify the most pressing water-related concerns, potential opportunities, and desired outcomes and indicators of amelioration. We came up with a list of ten strategic priorities for Canada:

- Drinking water for rural communities
- Water for nature
- Water efficiency and conservation labeling

- Wastewater management
- Canada water week
- Telling the Canadian water story
- Economic value of water
- Web-based information site about Canadian water issues
- Shared decision making
- Trans-boundary water governance

One interesting feature of this list is that almost a third of the priorities identified by my fellow WPAC members and I involve public education about the state of Canada's water resources (i.e. #s 5, 6 and 8). And that brings us to the pressing need for the present book, *Canada's Water, Yours to Protect: A Primer on Planning Together*, by Sandra E. Smith.

As Sandra Smith explains in the Introduction, hers is the latest in a series of books to address challenges and initiatives related to Canada's most valuable resource. What is particularly useful in the present book is its emphasis on not only identifying and reviewing the litany of problems but also in introducing and discussing a suite of corrective measures that have been undertaken (to varying levels of effectiveness). Another strength to the book is that it is authored by a single individual, thus bringing a unified, objective, and non-polemical, tempered and expert voice to the subject. And it is certainly a voice that those concerned about the state of Canada's water resources need to hear. For Sandra Smith comes from a background well immersed in the complexities of integrated water management due to her experience in the frontline trenches of professional practice. Much of that professional practice involved the discipline of planning, in my mind one of the most useful ways to address water management problems (France 2002, 2006, 2008). The planning emphasis is shown in the subtitle and penultimate chapter in the book. Also, Dr. Smith's time spent in academia educating about water management issues rings forth in the clear pedagogy found on every page. As such, *Canada's Water, Yours to Protect* is an important addition to the corpus of Canadian water management publications.

The topics covered in this book: water availability (surface and below ground sources), public safety and water's value (aquatic integrity, drinking water, flooding), and water development and infrastructure (dams and export) and planning, address all the remaining strategic Canadian priorities flagged by my fellow WPAC members and I in the list above (i.e. #s 1, 2, 3, 4, 7, 9 and 10). Finally, Smith's message pitched in the Introduction and revisited again in the final chapter is that Canada's water woes are not abstract, distant problems. Rather they are concrete, fixable challenges that require direct involvement by stakeholders—in other words, us. Hence the secondary clause of the main title: "**Yours** to Protect". As Abbott (2008) rightly stated, effective water management comes about only when approaches are "grounded in people, not technology, and in practice, not abstract theory." In this regard, Sandra Smith's book is a useful reminder that in terms of water management "people need to feel ownership of not only the problems but also the solutions" (France 2006).

—Robert L. France
Professor of Watershed Management,
Department of Environmental Sciences
Faculty of Agriculture, Dalhousie University
Principle,
W.D.N.R.G. Limnetics
Member,
Water Partner Advisory Committee
Council of the Federation of Canada

References

Abbott, R.M. 2008. "Visible Cities: A Meditation on Civic Engagement for Urban Sustainability and Landscape Regeneration," Foreword in France, R.L. (Ed.) *Handbook of Regenerative Landscape Design*. Boca Raton: CRC Press

Anonymous. 1997. *Rivers of Canada: How They Shape our Country*. User's guide book and CD-ROM. Harcourt Brace Canada

Armstrong, C., M. Evenden and H. V. Nelles. 2010. *The River Returns: An Environmental History of the Bow*. Montréal: McGill-Queens

Benidickson, J. 1997. *Idleness, Water, and a Canoe: Reflections on Paddling for Pleasure*. Toronto: University of Toronto

Brooymans, H. 2011. *Water in Canada: A Resource in Crisis*. Edmonton: Lone Pine

Casey, A. 2009. *Lakeland: Journeys into the Soul of Canada*. Vancouver, BC: Greystone

Campbell, C.E. 2005. *Shaped by the West Wind: Nature and History in Georgian Bay*. Vancouver, BC:

University of BC

Council of the Federation. 2010. *COF Water Charter*. Council of the Federation

—2011a. *Council of the Federation Water Stewardship Council: Terms of Reference*. Council of the Federation

—2011b. *Council of the Federation Water Stewardship Council: Water Partner Advisory Committee. Terms of Reference*. Council of the Federation

Evenden, M.D. 2007. *Fish versus Power: An Environmental History of the Fraser River*. Cambridge: Cambridge University

France, R.L. (Ed.) 2002. *Handbook of Water Sensitive Planning and Design*. Boca Raton: CRC– Lewis

France, R.L. 2003. *Deep Immersion: The Experience of Water*. Winnipeg: Green Frigate Books

France, R.L. 2006. *Introduction to Watershed Development: Understanding and Managing the Impacts of Sprawl*. Lanham, MD: Rowan & Littlefield

France, R.L. 2009. "Sprawlscapes and Timelines," in Shearer, A. et al. (Eds.) *Land Use Scenarios: Environmental Consequences of Development*. Boca Raton: CRC Press

France, R.L. 2013. "Exploring the Bonds and Boundaries of Water Management in a Global Context," pp. 1-3 in *Journal of Cleaner Production*: 60/1

Harrison, J. 2013. *A Timeless Place: The Ontario Cottage*. Vancouver, BC: University of BC

Huck, B. et al. 2002. *Exploring the Fur Trade Routes of North America*. Winnipeg: Heartland

Innis, H. 1962. *The Fur Trade in Canada*. Toronto: University of Toronto

Jennings, J., B.W. Hodgins, and D. Small. (Eds.) 1999. *The Canoe in Canadian Cultures*.

Toronto: Natural Heritage Books

Mihell, C. 2012. *The Greatest Lake: Stories from Lake Superior's North Shore*. Toronto: Dundurn

Morse, E. 1969. *Fur Trade Routes in Canada, Then and Now*. Toronto: University of Toronto; Ottawa: Parks Canada

Newman, P. 1986. *Company of Adventurers*. Toronto: Penguin

Newman, P. 1987. *Caesars of the Wilderness*. Toronto: Penguin

Raffan, J. 1999. *Bark, Skin and Cedar: Exploring the Canoe in Canadian Experience*. Harper/Perennial

Robertson, H. and M. McCracken. 2003. *Magical, Mysterious Lake of the Woods*. Winnipeg: Heartland.

Russell, F. 2004. *Mistehay Sakahegan, the Great Lake: The Beauty and the Treachery of Lake Winnipeg*. Winnipeg: Heartland.

Unwin, P. 2003. *The Wolf's Head: Writing Lake Superior*. Toronto: Cormorant Books

PROLOGUE
A PERSONAL JOURNEY

Canada's Water, Yours to Protect: A Primer on Planning Together comes from my practical and academic experience in water management and land use planning. Because of this, my voice varies from policy-maker to teacher to the personal. A short story and a career path description explain. On my second day working with British Columbia's water management program, I attended a retirement party for a man whose career there had extended over fifty years. In a conversational gambit that could disqualify me for future promotion, I asked the Director how anyone could work on "water" for that long. This kind man smiled mysteriously and said: "When you know the name of every watercourse in British Columbia, then you will understand."

His response to me began a journey that included 17 years in water management, mainly developing water policy, and 4 years in island planning. Co-authoring *Domicide: The Global Destruction of Home* (Porteous and Smith 2001) extended my interest in the effect of top-down planning and emerging participatory approaches. Then I added to my experience by teaching water management at an undergraduate level. Alfred Stieglitz wrote to Ansel Adams: "I chose my road years ago - & my road has become a jealous guardian of me. That's all there is to it" (Adams 1984, 69 cited in John Updike's *Just Looking: Essays on Art*). *Canada's Water, Yours to Protect* is one more step on my path, seemingly both chosen and chosen for me. But this book is not my story alone. I am grateful to many teachers, some named within, and to countless others: water managers, planners, politicians, academics, journalists, community members and students, who have enriched my journey, and hence my voice.

CHAPTER 1: INTRODUCTION

Canada's water is part gift, albeit capricious; part history (our second nature); and part challenge. Think of Walkerton, Ontario, where seven people died as a result of contaminated water in 2000; Tofino, BC, which experienced a serious drought in the summer of 2006 despite an annual rainfall of 324 centimetres; and predicted damage to the Athabasca River watershed by the oil sands industry. Warnings about the world's water amplify these concerns.

Reports from the World Business Council on Sustainable Development, the United Nations Environment Programme and the International Water Management Institute present the evidence. These organizations asked 700 agricultural research scientists, 1500 experts on the social effect of water scarcity in developing countries, and executives from the world's largest water, oil and chemical companies to forecast the future of the world's water. Their verdict came "with alarming unanimity. Each survey independently predicted supply, health or economic crises coming sooner rather than later if there is no radical change in the way water is used" (Vidal 2006, 1). In 2009, UN Water suggested that conditions "may worsen, converging into a global water crisis and leading to political insecurity and conflict at various levels" (UN Water 2009, 2).

What is the end-point? Looking back from the year 3000 in her futuristic article, Margaret Atwood remembers when "hydroelectric power failed because rainfall became so infrequent; and California depleted its aquifers and became a salinated wasteland, and then there were the Water Wars of the early 2020s, which drained the Great Lakes ... " (Atwood 2000, M4). Water organizations, academic sources and popular writing cite threats to the ecosystem, health, economic activity, social customs related to water, and the intrinsic values accorded to water. Can such alarming predictions be forestalled?

Canada's Water, Yours to Protect

Many respected sources speak of the problems besetting Canada's water and propose solutions[1]. *Canada's Water, Yours to Protect* joins this dialogue, but takes a different approach. While acknowledging the often essential role of government, the effect of local collaboration driven by passion for place is celebrated. This book encourages Canadians to come together to plan for the future of their water. Meaningful and locally responsive input to governance is one desired result; enhanced involvement in volunteer and public interest group activities is a second.

To give context and help achieve these goals, the present book explains who does what and why, and what the future challenges are for eight issues we now face:

- deeply-entrenched water allocation systems
- ill-protected groundwater supplies
- full-cost water pricing versus water conservation
- aquatic ecosystem protection
- drinking water quality protection
- safety from flood damages
- future dam construction, and
- control of water export.

Then a chapter on planning and integrated management illustrates the benefits of comprehensive action and community input relating to each of these areas. As this book shows, in addition to traditional government sources, every community and every individual, while using and enjoying water, must ensure its protection. Seeking sustainability[2] of Canada's water is the ultimate goal; focused governance, greater social responsibility, and an enhanced "water ethic", the desired future.

Canada's Water, Yours to Protect is about Canadian initiatives and challenges relating to fresh water. Written for multiple audiences, it reflects multiple constituencies who hold the keys to better water management. General readers and community stakeholders want to understand how to alleviate water concerns and what they can do themselves. Students from many disciplines choose to learn more about water, particularly those in geography, history, planning, law, political science, sociology, environmental studies, hydraulic engineering and water-resources management. So do politicians, government policy-makers, the private sector and interest or volunteer-based organizations and the media.

My focus is on Canadian experience and relies almost exclusively on sources from across Canada. But as British Columbia is my home province, it features frequently. Many books about water are readers or university texts, authored by Canadian academics and lawyers who are water experts.[3] As a single author, I hope to offer a coherent approach, the benefit of practical and academic experience and a balanced take on issues which are often approached polemically. My major focus is planning for water, but I try to be comprehensive in the treatment of problem areas. For example, I include chapters on floodplain management and the construction of dams, two issues of increasing public importance. I have written an extended case study on the protection of groundwater by an island community, focus on the media's attention to water issues, and include my approach to speaking for the intrinsic values of water.

How Did We Get to Here?

Governments now play a dominant role in how we manage water (Chapter 2), but to begin to understand how thinking about water has developed, we need to look to the past—here, through a geographer's lens. Exploration to collect and classify information about "unknown" parts of the earth paved the way for future settlements and new economic riches. Exploration by boat played no small part; water as transporter. Margaret Ormsby describes Mackenzie's travels on the Fraser River: "Tremendous hazards lay ahead. On June 19, [1793] he embarked his party at three o'clock in the morning; ... A quarter of a mile along, the exhausted men attempted to run successive cascades, in a very turbid current, and full of whirlpools, their canoe filled, and they reached the bank in a half-drowned condition" (Ormsby 1958, 31-32). Much later, exploration gives way to adventure tourism as for example, Clara Vyvyan recorded of her journey on Manitoba's Rat River in 1926: "More than once we came to what the guides termed "swift water" and we regarded as rapids, and several times, in order to reach an easier shore, we were ferried across one of these rapids on a diagonal course, swiftly losing ground as the current carried us downriver at a giddy pace" (Vyvyan 1998, 102).

From the 1880s, public health concerns began the long path to the treatment of water quality we see today. In the early 1900s, environmental determinists saw nature (particularly climate) as responsible for patterns of human activity on the earth's surface. Perceptions of water or lack thereof, as a threat led water managers to hold a "command and control" ethos. This led to construction of stream diversion works, dams, canals and ports (Rosen and Tarr 1994, 302). Many other approaches evolved in the last fifty years.

From the 1960s, the search for laws about natural and human phenomena in space has influenced thinking about ecosystems and possibly formed the basis for water management planning (Johnston 1983, 115-117); water as integrator. Developments in resource management brought "waves of innovation" to deal with the increasing complexity of water problems (Dorcey 2004, 531-32): planning, environmental assessment and new public involvement methods. Since the mid-1980s, a number of high-profile environmental disasters justified greater regulation (Abbott 2009, 10). Sustainability concerns fostered multi-stakeholder involvement, conflict resolution and consensus building. Dorcey (2004, 532) says that while useful, these techniques were seen as time-consuming and costly during a weak economy. A new century brought a new wave: "revitalizing democratic governance processes" and increasing interest by the media. Present day economic concerns foster cuts to funding for both government and interest groups and streamlining of regulations.[4] Day-to-day practice managing water, combined with academic studies and public involvement, will continue to influence the way we think about water and are the subject of this book.

Organization of the Book

Canada's Water, Yours to Protect contains twelve chapters, eight of which look at specific issues in water management. More general themes such as climate change and water sustainability might have formed separate chapters; instead these issues appear throughout.[5] Chapter 2 is a brief introduction to water management, particularly for those new to the subject. It speaks of Canada's fresh water supplies, the issues, the actors and the changing role of public involvement. Beginning the theme of water availability, Chapter 3 examines the practices used to divide surface water between various demands and the conflicts that arise. These water allocation decisions underlie sustainability issues discussed in later chapters. Continuing this theme and adding to the later chapter on drinking water quality, Chapter 4 is mainly the story of one coastal island's community activism to ensure its future supply of groundwater. Also related to availability, Chapter 5 discusses the economics of water and water demand management. Water pricing is an essential government tool, but public and private sector efficiency and conservation measures are often considered more desirable.

The themes of ecosystem protection and public health and safety begin in Chapter 6 with an emphasis on local government and volunteer efforts to protect the aquatic ecosystem, and second, on initiatives to preserve intrinsic values of water. Central to this book's argument, the chapter highlights past efforts and new methods to compensate for insufficient senior government protection. Chapter 7 is about protecting drinking water quality, synthesizing Canadian response and hoping to build a knowledge platform for new players. Chapter 8 discusses flooding and whether current approaches are adequate to protect Canadian communities, a question many residents are asking after the floods of recent years.

Chapter 9, part of the third theme (water development and infrastructure), seeks a balanced view about dams and highlights the need for involvement by those most affected. Chapter 10 examines water conflicts - the thorny issues of bulk water export and bottled water - areas of interest heavily motivated by media and public influence. At this point, in Chapter 11, the emphasis changes to show how the issues come together in water-related planning and integrated resource management, informed by strategic approaches and public collaboration. The concluding chapter proposes new directions. Considered together, these chapters hope to broaden the reader's perspective on water issues.

The search for better ways to manage water is increasingly relevant. Success in these efforts will maintain economic security and bring us the goods we consume; social security including protection from floods; and ethical security, upholding the right to water for people and other species (Acreman 2001, 258): water as a basis for life. Today, this search for sustainability grapples with climate change, new economic realities, and governments more interested in development than the environment. Against such odds, we need to re-invigorate the best of past practices and find new, more strategic, approaches; even though success may be achieved in small increments. We need appropriate local-level governance, creative planning approaches and more inclusive decision-making; greater social responsibility and commitment shown through public awareness and engagement on water issues by both actors and influencers; and adoption of an enhanced "water ethic".

CHAPTER 2: BEING INVOLVED

Chapter 2 defines water management, briefly describes Canada's water resources and explains how issues arise and how actors/influencers[6] react. Attracting the ideas and actions of others to formal water governance is an important focus here and the subject of the final section (see also Chapter 11).

What is Water Management?

On the surface, water management seems to mean just managing water, but it is more about managing the human use of water. "Water management is inherently a question of governance. Water related issues ripple throughout society and affect basic livelihoods and deeply embedded social values. As a result they must be addressed at a societal level through the complex array of political, economic, institutional and social processes by which society governs itself" (Moench 1999, 2). Important questions arise: Should policies operate to increase the government income or to improve the quality of the environment? Which water uses should have the highest priority: drinking water provision or power development or irrigation or fisheries protection? Who should be part of the decision-making process? Water management affects how we use, protect and develop water resources.

Yet water management is much more complex than a simple definition implies. Its players follow diverse strategies: watershed and land use planning that affect the availability and quality of water; water supply management (withdrawal, pre-treatment and distribution of water from surface or groundwater sources); water demand management seeking the efficient distribution and use of water; water quality management to return clean water to the environment; and remedial strategies, including emergency response and long-term clean-up or ecosystem restoration (Shrubsole 2001, 1). Other strategies exist for floodplain management and for the creation and operation of dams.

Water management has a long history[7], but in essence, it changes over time, influenced by external events. Climate and hydrology alter a place or ecosystem and the result interacts with human activities. In turn, these activities affect both land and water (Figure 2-1). Issues and conflicts arise and various actors/influencers try to resolve them. The easiest way to understand this is to ask Canadians to tell their water stories. I can only guess at the richness of their response. I hope that they would start by telling you of a river or lake experienced in their childhood or explored in their travels, about the meaning of water to their way of life, about swimming or boating or fishing or just pausing in sheer wonder. If they live in a rural community, they might speak of the privilege of having a water licence or permit, being a riparian owner of water rights or drawing water from their own well. But they might also tell you of the frustration of not being able to get water given drought or overuse/over-allocation of a scarce resource. If they live in an urban area, their stories could be of watering restrictions, boil water advisories, increasing rates for water or waste disposal, or more positive, the value of a cherished water view.

Too often their stories may be of contamination of a water source or of a severe storm or rapid runoff that flooded homes and fields. Some will tell how they fought against the construction of a dam, use of their water for bottling, or the export of Canadian water. But there will also be stories of working to collect information about their water source and its value to them, the importance of water to a business venture, learning about water, learning how to conserve water, how they built a dam or dike or reservoir, restored a stream or wetland or were involved in a planning process about the future of their water. There are many stories and many perspectives.

When issues arise, diverse actors/influencers then undertake/shape water management activities. Elements of this framework and their foci change over time (Figure 2-2 —page 20). As new issues arise; new voices propound theories and ideas. Below, a discussion of Canada's water resources and a brief introduction to the issues and actors/influencers explain this and act as background for subsequent chapters.

Water: A Short Description of a Critical Resource

Water is a crucial, non-substitutable resource. About 70 percent of the human body is water; how long could you live without it? The human body needs 2 litres of water a day and you would last only a few days, facing death once water loss exceeded 10 percent of body weight. Water is also important in itself, as part of the environment, and as a basis for settlement and economic development. Of the entire world's water supply though, just over 2.5 percent is fresh water and most of this is in the form

Figure 2-1: Water Management Framework

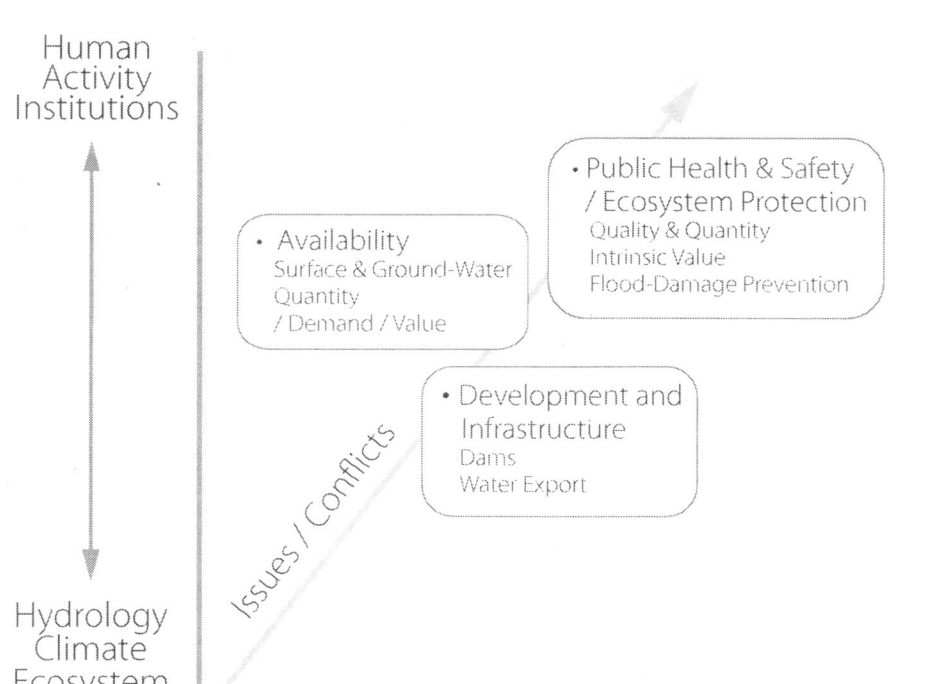

of ice or groundwater; only .01 percent is surface water and more easily accessible (Canada Environment 2004c). While this is not a book about hydrology,[8] the operation of the hydrological cycle as water changes its state and composition and is distributed, is "as crucial for life on Earth as is the presence of liquid water in the first place" (Ball 2001, 25). Canada's apparent good fortune illustrates the pattern of water's existence.

Canada has about 6.4 percent of the world's renewable fresh water (much less than Brazil, slightly more than the United States (Canada Environment 2010b; Pacific Institute 2006)), but this estimate misleads. Precipitation can be extremely heavy in coastal areas yet decline rapidly inland to less than 50 centimetres per year, even as low as 12.5 centimetres (Smith 2001, 66). Southern Canada, where 98 percent of the population live, is responsible for 38 percent of the water yield (renewable fresh water); in the North, water yield per capita is 185 times greater

(Canada Statistics Canada 2010, 5).[9] Water volume also varies considerably over time with differences from day to day, from season to season, and cyclically. The result can mean droughts and floods, not simple abundance. This uncertainty is growing due to climate change. From ocean to ocean to ocean, there is evidence: decreasing glacial runoff and rising lake levels in the circumpolar region, flooding on the Yukon coast, greater spring flood damage and low flows due to drought in the West, and increased droughts and flood flows for the Prairie provinces. Researchers observe falling water levels on the Great Lakes and the St. Lawrence River, storm surges causing erosion and flooding and low water levels due to drought in the Atlantic provinces. From 1971 to 2004, water yield in southern Canada decreased 8.5 percent, an average annual decrease almost equal to the amount of water supplied to Canadian homes in a year (Canada Statistics, 7).

Figure 2-2: Change in Emphasis over Time

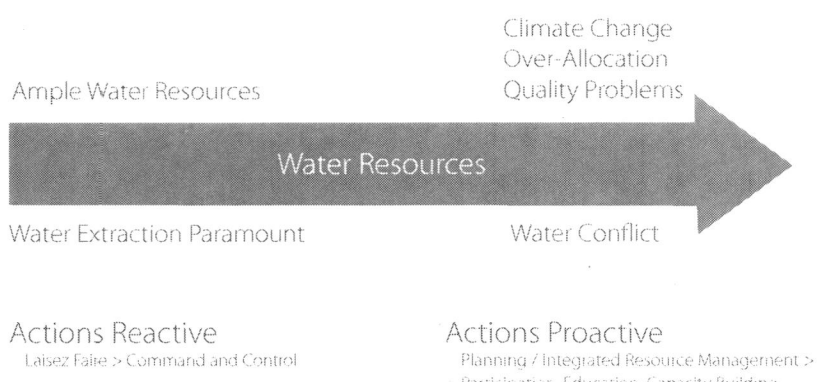

Ample Water Resources

Climate Change
Over-Allocation
Quality Problems

Water Resources

Water Extraction Paramount

Water Conflict

Actions Reactive
Laisez Faire > Command and Control

Actions Proactive
Planning / Integrated Resource Management > Participation, Education, Capacity Building

Incremental Actors / Influencers
Individuals/User Groups/Public Health Officials/Politicians/Government Officials/Infrastructure Builders and Operators > Lawyers / Engineers /Integrated Resource Managers / Planners /Multi-disciplinary (Scientists and Social Scientists) > Public Health Specialists/Public and Private Water Supply Operators/Media > Community/Special Interest Groups/Stewardship Groups/Volunteers

Issues and Actors/Influencers

Water supplies many demands: drinking water essential to our survival; flows giving habitat for aquatic ecosystems, navigation as part of transportation, recreational pleasures, and aesthetic delight; and surface water and groundwater, enabling riches. Hydrological setting and climate can influence each of these uses. Problems arise when individuals or uses desire the same scarce water, pollute that water, or develop projects that deleteriously affect the resource. Actors or influencers working at multiple scales address these challenges.

Government institutions predominate, but many others play significant roles. Together, they collect information, practice and propose policy in response to a series of perceived issues and conflicts, integrate indigenous knowledge/justice, engage stakeholders and build citizen capacity. They undertake interdisciplinary research and create knowledge and policy capacity. They influence change through academic studies, volunteer work, public interest advocacy, and media intervention.

For much of the 20th century, non-political decision-makers in water management came from the technical elite, principally engineers. Hydraulic and civil engineers build the works that control water, but their professional activities have broadened to include environmental assessment and planning. Their work was and is supported by geomorphologists who examine natural features of the earth's surface and the processes that change them. Hydrologists study the occurrence and movement of water on and over the surface of the earth and provide information about precipitation, evaporation and transpiration, stream flow, and groundwater flow. Surveyors record the shape of bodies of water and their soundings reveal the depths.

Water management has also become the prerogative of economists, geographers, agronomists, lawyers, sociologists, philosophers, political scientists, historians, management specialists, resource managers, aquatic scientists, and public health experts. Planners enable the orderly disposition of land, resources, facilities and services in urban and rural areas, while integrated resource managers ensure sustainable use, management and development of water, land and related resources—their integrative and collaborative talents essential to the resolution of increasing levels of conflict and the adoption of a proactive approach. Today, we need multidisciplinary knowledge and skills: legal, economic, technical, operational, and planning, as well as communication and business techniques. And with these, we need the special knowledge and skills of stakeholders, interest groups and individuals.

Government

The political arm of government is at the forefront of decision-making and is quickly assigned responsibility when things go wrong: "The Walkerton inquiry into the illnesses and deaths caused by *E. coli* contamination of the town's water supply heard appalling testimony yesterday and the day before about just how badly its Public Utilities Commission was managed. That mismanagement was so extreme, and went on so long without being caught, that chief commission counsel Paul Cavaluzzo declared, 'I think Ontarians should seriously consider, is this happening across Ontario?' The Mike Harris government can't promise it isn't" (Ibbitson 2000, A7). But there is a more positive view.

You can find many excellent sources of information on water governance (Appendix 1 gives a brief summary).[10] In general the focus of the public service is now more proactive than reactive. "In a world in which "government" is replaced by "governance", governments are increasingly being judged on their ability to manage complex relationships and to work across sectors, bridging what sometimes can be intractable cultural and organizational divides" (LaForest and Orsini 2005, 484).[11] Historically a service delivered by individual government agencies, water management has evolved to delivery by partnerships. Driven by economic and ideological factors, the federal and provincial governments download responsibilities, often to the local or volunteer level. The focus is on planning and integrated resource management, adaptive management,[12] citizen collaboration, and education, rather than "command and control" activities.

Partnerships

Partnerships make resources stretch, encourage compromise, allow for better coordination, and build up legitimacy for government organizations (Haughton 1998, 15). Two specific examples illustrate. Partners FOR the Saskatchewan River Basin promotes water stewardship by developing information and education programs.[13] They have delivered 20 major projects since 1993 (Partners FOR the Saskatchewan River Basin 2011, 1). The Canadian Water Network, established in 2001, links academics, private industry, government and not-for profit organizations to "mobilize multidisciplinary research excellence, catalyzing innovation that provides practical, implementable solutions to complex water resource management issues" (Canadian Water Network 2011).

The private sector including agriculture (discussed in later chapters) is an important partner in water management. The Royal Bank of Canada Blue Water Project leads the way with its ten-year, $50 million donations program for organizations involved in water protection. In 2012, the Project gave $2 million to a Queen's University Water Initiative to enable students to work with industry on water problems. This funding will also allow local school students to experience a watershed protection process. The private sector, often assisted by government also develops technological solutions to water problems such as pumps, valves, filters, and waste-water treatment systems.[14] Other companies, such as consulting engineering firms, provide management (sometimes design and construction) of water systems in Canada and internationally.

Early partnership examples are water supply utilities, irrigation districts, conservation districts, and improvement districts. Large scale private-public partnerships (frequently referred to as P3s) that provide water and treat wastewater have recently become more common in Canada and as such, are the subject of critical comment (for example, Bakker (Ed.) 2007, 185-204; Barlow and Clarke 2003; Brubaker 2002). A positive view sees a public-private partnership as "a co-operative venture for the provision of infrastructure or services, built on the expertise of each partner, that best meets clearly defined public needs, through the most appropriate allocation of resources, risks, and rewards ... the public sector maintains an oversight and quality assessment role while the private sector is more closely involved in actual delivery of the service or project" (Canada Industry (2003, 1).

There are also problems that may outweigh the advantages of such arrangements. Early criticism outlined the disadvantages associated with public-private partnerships: "loss of local government or utility control, unfair competitive process due to bias in selection process, opposition by unions, middle management and community, threat of bankruptcy by private company, deferred maintenance, and poor private company performance" (Moraru-de Loë (1998). Add to these, "... confused lines of accountability, increased user costs, loss of public sector jobs, limited competition, limited control over public policy, ... transfer of assets, leaking of confidential information, ... and [less concern] with social responsibility, environmental issues, local knowledge" (Canada Mortgage and Housing Corporation 1999 cited in Canada Infrastructure 2004b, 13).

Good governance and extreme vigilance are central to avoiding the pitfalls: "[this] requires a regulatory framework that covers water quality, environmental issues, and economic issues" and a "municipal government [with] access to good information and resources sufficient to carry out its oversight duties" (Bakker 2003, 37, 38). Also important are public scrutiny, on-going systematic project evaluation, an independent regulator, and consideration of all options (Ouyahia 2006, 3).

One estimate suggests that this type of arrangement for municipal water supply serves about 5 percent of the world population (Stephenson 2005, 265 cited in Ouyahia 2006, 13). Canadian examples are less common. Ontario's Clean Water Agency was formed in 1993 to manage water and sewage facilities at the request of municipalities. In 1996, Ontario's Minister of Environment announced a plan to get out of the water business through privatization, but this has not proceeded and in 2011, the Agency still serves close to 180 clients (Ontario Clean Water Agency 2010). For Canadian investors, publicly-traded companies that build and operate water systems are of growing interest. In 2006, the Canadian Pension Plan Investment Board, as part of a consortium, made an offer of $4.7 billion (Canadian) for the parent company of Anglian Water (that serves around 6 million customers in the east of England) (Grant 2006, B19).

Other Players

Other influencers have a profound effect on water management: professional organizations, special interest and volunteer groups, the academic community, the media, and individual members of the public. Each pursues its own agenda and, as seen later in this chapter, each reacts to opportunities created by government and others. As Savoie observes: "All public institutions in Anglo-American democracies have been told in recent years to reach out to client groups, interested parties, research institutes, and NGOs when shaping public policies" (2010, 15).

Professional organizations exchange views and promote change, as two examples show. The Canadian Water Resources Association has regional branches that include managers, administrators, scientists, academics, students and users among their members. Conferences and publications (*Canadian Water Resources Journal* and a newsletter, *Water News*), update members. The Canadian Water and Waste Association represents public sector municipal water and wastewater services and their private sector suppliers and partners. Assisted by regional associations, this organization holds conferences and has technical committees in particular interest areas (Canadian Water and Wastewater Association 2012).

Not for profit, non-government organizations have a long history of influencing environmental decisions through research and project funding. Examples include the Walter and Duncan Gordon Foundation that funds public policy development relating to the Canadian North, global citizenship, and fresh water resources. The Sierra Legal Defence Fund (now Ecojustice) intervenes in environmental issues. Their report, *Waterproof* 2001 (Christensen 2001), showed that Canadian provinces and territories had inadequate laws to ensure the safety of drinking water. Alberta WaterSMART promotes open sharing of water information, a program for new water technologies and practices, an Awareness Building Program, and a Water Entrepreneur and Technology Centre. Smaller volunteer groups lobby for change and spearhead conservation efforts (see Chapter 6). One example shows just how useful their work can be. The Lake Ontario Waterkeepers produce primers including

"Taking water from the Great Lakes: a citizen's guide to the policies, rules, and procedures that protect Ontario's waterways" (Lake Ontario Waterkeepers 2006).

Water is of considerable academic interest in Canada. Some universities pool academic resources from many departments to form groups such as *The Water Institute* at the University of Waterloo. Others have a single department focus on water such as the Water Governance Program at the University of British Columbia. The call for relevance, for partnerships with government and community, and for research funding affects most academic departments. Recent work relating to water quality, privatization, climate change, community capacity, and the political aspects of water management is evidence of this trend. Also producing critical work are university-based think tanks such as the POLIS Project on Ecological Governance at the University of Victoria, the Adaptation to Climate Change team at Simon Fraser University and the Munk Centre Program on Water Issues at the University of Toronto.

External to all the sources described above are the media. Water problems supply continuous fodder: "An untimely union of drought and development in the small beach town of Tofino has put many long-awaited summer plans on hold—including ... [a] wedding. Their Sunday afternoon service on the Wikaninnish Inn's waterfront property has been cancelled, following the district's request to shut down businesses as of tomorrow, an unprecedented water-preserving measure" (Salinas 2006, S1). References to the media, particularly the national newspaper, *The Globe and Mail*, and the Canadian Broadcasting Corporation, occur throughout the present book. These sources are just the tip of the iceberg of a now pervasive television, radio, print, and traditionally unconventional internet-based (including the blogosphere's citizen journalism) presence. While we may sometimes feel that, "the media rewards hyperventilation" (Hébert 2009), review of these sources gives topical insights.

The media pay attention to water issues, bringing forward urgent matters, seeming a "more important component of the political process than parties and electoral systems" (Hobsbawm 1994, 581). The media try to be at the source of issues first, sometimes using this advantage to reveal controversies. Their skill in information-gathering, story-telling and analysis is evident and their reports are often easier to understand than government sources; they may act as transmitter and transformer. When an environmental journalist becomes an author, the results provide specific insight; Hanneke Brooymans (2011) defines the state of Canada's water as a crisis. The media influence water management in three ways: their articles engage public interest so affecting the importance of an issue; they may set the agenda for government considerations, even paving the way for regulatory regimes; and they have the potential to influence behavioural change (Boykoff 2008, 566).

As discussed in later chapters, volunteer organizations and groups influence water management as they participate in protection and restoration activities. Individuals contribute their specific interests. A little anecdotal evidence illustrates the motivation and perspective of Jane Henderson as she tried to protect her community's drinking water. Jane's story relates to two local governments (Moncton, New Brunswick, and Cochabamba, Bolivia, and their surrounding areas) and two multi-national companies (USFilter Canada and Bechtel).[15]

For years the drinking water supply of Moncton, Riverview and Dieppe (Moncton) was turbid and discoloured, with a bad taste and smell (Bertels and Vredenberg 2004, 43). New provincial standards and repeated boil water orders meant that the water treatment plant must be improved. Moncton considered a user-pay model, but this did not pass, given other priorities. Provincial or federal funding was not available. Moncton issued a tender to the private sector and USFilter was the successful bidder to design, build and operate the plant. The project improved the water quality; Greater Moncton Water Ltd. (a public-private partnership) became the new manager. The City of Moncton is responsible for protecting the water source, transmission and distribution lines while USFilter, and ultimately Veolia Water Canada, (a branch of Veolia North America) manages the Greater Moncton Water Treatment Facility.

In 2001 Jane was a student at Mount Allison University in Sackville, New Brunswick. A major in English, obsessed with John Donne, she nevertheless joined a public lobby group that had water supply concerns. It wasn't even her water supply. She says: "But that's the point. What qualifications should I need aside from being a person who lives in the 21st century? When companies start buying up water, that affects me; it's not like people who aren't experts aren't connected to these issues." Jane took the opportunity to attend a People's Global Action conference in Cochabamba about anti-globalization and to discuss the Moncton experience.

It was there that she learned more of a now well-known story; in April 2000, the city of Cochabamba was shut down for 3 days during a conflict over supply of their water through a contract with a private-public partnership, Aguas del Tunari, of which Bechtel, a California company, owns 27.5 percent.[16] She learned how the water rates went up as much as 200 percent and how the government took control of groundwater and new wells. There is much more to this story: the vagaries of trade agreements and Cochabamba's eventual adoption of a publicly-controlled water supply. But, as for Jane, she returned to Moncton and told the local Labour Council of her experiences. Invited to appear in the audience of the CBC program *Counterspin*, just as the credits rolled, she had the chance to ask the Mayor of Moncton: "We've heard about accountability and transparency, but what about irretrievability?" Jane's question came from her knowledge about the North American Free Trade Agreement (NAFTA) signed by the governments of Canada, the United

States and Mexico (discussed in more detail in Chapter Ten). Jane continued her involvement with the issue through SOS-EAU-WATER-SANKWAN and other interested groups.

Many factors drive individuals' involvement in water management. Jane's motivation is the fear that "[w]ith globalization, the free market, and the search for international competitiveness and economic efficiency, benefits will accrue to the few, and the hopes of the many will be dashed" (Porteous and Smith 2001, 240). Jane would be pleased that a Water Action Plan (Moncton 2008) now exists and that in 2010 Moncton received $20 million from Canada's Economic Action Plan to ensure an adequate water supply for current and future development needs. She would be less pleased with the delivery of water treatment by the private-public partnership.

Which of these many players, inside or outside of government, are in the best position to manage water? The following chapters, divided into sections by the themes of water availability, protection and development, show the roles of each. Senior governments are essential given the need for laws and funding. Local government often makes the greatest difference. Citizen collaboration brings inestimable benefits. But all, all of the players are necessary if future improvements are to occur, particularly given new economic realities. We need to find ways to encourage opportunities for everyone's input.

Opportunities for Involvement

Consultation with and participation by stakeholders and the public in water management are not new (see Goodman 1984, 188-193; Diduck 2004).[17] The purpose is often to receive comment on proposed policies and legislation. Preparation of the Canadian federal government's 1987 water policy included an inquiry, preparation of expert papers, and consultation within and outside government. The International Joint Commission used more active public involvement in their work on the water export issue as did the federal government in work on Action Plans (discussed in later chapters). The provincial and territorial governments encourage input to discussion papers and reports through public meetings, inquiries and study groups. Often, the emphasis is on education and, to some degree on placation, particularly post-Walkerton.

At a local level, earliest efforts targeted water users and those most affected by large projects; again mainly through education, rather than consultation. One exception is the appointment of "advocacy" planners to assist the "planned for" who need to make sense of project plans and complicated bureaucratic language. By articulating stakeholders' interests, more acceptable alternatives might be achieved. An advocacy planner assisted the community during the development of the

Revelstoke Dam project in BC, but in this and other circumstances, the expectations placed on one individual are often overwhelming. A second exception is where the community takes the situation into its own hands. "People need to feel ownership of not only the problems but also the solutions..." (France 2006, 21). Residents of Ontario's Tiny Township and natives from a nearby reserve blockaded access to a construction site. They feared that a landfill would contaminate their aquifer that supplies uniquely pure water (Mittelstaedt 2009, A3). Ultimately, their protest has been successful.

Why do agencies involve the public and what do they hope to achieve? It is more than thirty years since Sherry Arnstein wrote her seminal article on citizen participation. Her context is urban rather than environmental, yet her message is still relevant. For Arnstein (1969), citizen participation is another term for power. She sees participation efforts arranged on the rungs of a ladder—each step corresponding to the extent of a citizen's power to influence a plan or program. At the two lowest levels, the public is subject to a form of manipulation and therapy where persons in power "educate" or "cure" the participants. At rungs three, four, and five, citizens hear or are heard, but cannot ensure that their views will make a difference as they are informed, consulted and placated. At the upper rungs, people are able to negotiate with persons in power and may, at the highest level, hold decision-making seats or managerial power.

Participation efforts continue to fall within Arnstein's framework. Today, these efforts may include "diagramming ... , mental and social mapping and modelling, transects and historical timelines, ranking and scoring preferences, observation, focus groups, and role play"; the goal is to seek truth by collecting and analysing information in the field, and allowing feedback to occur during the process (Hickey and Kolthari 2009, 86). Planning processes include "public hearings, public meetings, focus groups, and surveys as well as ... public representation on advisory boards, task forces, consensus building processes and collaborative work groups" (Laurian 2009, 381). Here interpreted in the context of plan preparation, three approaches offer varying opportunities for involvement (Shrubsole and Mitchell 1997, 312).

In the first, a public agency proposes a plan. Reviewers criticize or support it. The agency then defends or alters their plan and implements it. In the second approach, the agency seeks input before proposing the plan, but follows the process previously described. The third approach, consensus-based decision-making, achieves the higher rungs of Arnstein's ladder. Key participants (agencies, non-government organizations, other groups and the public) guide plan preparation. They express interests rather than positions, come to understand, respect and trust the interests of all, and ultimately adopt recommendations, even though for some, the position may not be their first choice. In these cases, facilitators record alternate views.

After plan preparation, all participants help implement the solutions—an infinitely preferable process: checking possible misuse of power, encouraging an informed electorate, and communicating attitudes of a diverse citizenry and the needs of a complex resource and its users. Additional goals are to provide values other than dollar values so encouraging balanced discussion; to give a nuanced and longer term understanding of the impact of policies on those most affected; to tap the expertise of each person, giving a sense of self-determination, while fulfilling the ethical imperative to understand their concerns; and to help resolve conflicts (Canadian Institute for Environmental Law and Policy 2004, 5; McAllister 1982, 240).

Despite important gains from public involvement, these processes take a long time, are costly and can go wrong. Problems can lead to siting of unwanted projects next to weak constituencies, loss of desirable outcomes when knowledgeable agencies do not participate, and failure to achieve essential political buy-in where leadership is not part of the outcome (Fainstein 2000). A 3-year ethnographic study of a public process undertaken about a water main extension illustrates another problem; participation comment by one interest group can dominate (Roth et al. 2004).

Persons with scientific and technical expertise interacted with those whose knowledge is local and long term. At an open house and a public meeting, technical and advisory bodies made presentations. The study's authors suggest that the scientists ignored the residents own cultural and historical understandings—the residents' sometimes emotional comments about living with substandard water contrasting with the scientists' cool objectivity. While the underlying argument in this case is whether extending the water pipeline would open the door to development in a partially rural area, a subtext of independent superior science versus concerned residents is very apparent.

How does the decision maker react in these circumstances and is it possible to resist past training that science proves all? Good public involvement may depend on, for politicians, not assuming that your values are a good indicator of the values held by the people you serve; for technical experts, not claiming to have special and superior values; and for the planner, ensuring respect for all values (adapted from McAllister 1982, 236). For a start, this means an initial process that passes no judgment during discussions nor allows coaching to get a response, ensures excellent facilitation, honours even the strangest suggestions for these often lead to the most creative solutions, and understands that the combination of thoughts will lead to the best direction, not each individual's thoughts.[18] Water management benefits from awareness-building, public participation, consultation and collaboration, but new initiatives, at either end of Arnstein's ladder, show new creativity. As part of the Ontario Children's Water Festivals, students from grades 2 to 5 participate in hands-on activities relating to groundwater: Royal Flush, No Water off a Duck's Back, and Filter Bed (Reid 2004, 3). Participation is a learning experience; although power isn't gained immediately at this rung, knowledge is likely to contribute to future opportunities to cause change.

A second example is asset-based community development. Dawn Carr, acting as a planner for the Canadian Heritage Rivers System portfolio for Alberta Parks and Protected Areas suggests that it "mobilizes" a community and helps them to identify their "assets and capacities, rather than needs, deficiencies and problems" (Carr 2004, 6).[19] Building community capacity relies on the idea that it is not only government that can give information, promote change and monitor the results. Building community capacity "enables the creation of social capital, the emergence of leaders and, through collective action, helps communities attract financial, human and technical resources that may last long after the issue at hand has been resolved or faded" (Phillips and Orsini 2002, 17).

I once heard John McKnight, Professor of Education and Social Policy and Co-Director of the Asset-Based Community Development Institute at Northwestern University, speak of the men from a community bowling club who realized that they had the capability to build a ball diamond for children. Then sharing a pride in their accomplishment, they wanted to take a continued interest in the improved facility and its weekly activities. So too, people whose knowledge, talents and skills create a vision for the future of a watershed might wish to help see their dream fulfilled.

Summary

This chapter introduces water management and those with the potential to shape the future of Canada's water. Water management is defined and a simple framework shows how issues arise and change over time. Canada's water resources and the various actors/influencers in water management are described. The change from "government" to "governance" means new approaches to stakeholder and public involvement: awareness building, consultation, citizen engagement, collaboration and the evolving nature of these activities. Together, we can create a new story that will ensure the sustainability of Canada's water.

THEME 1: WATER AVAILABILITY

CHAPTER 3:
ALLOCATING SURFACE WATER:

IS THERE A BETTER WAY?

> Water wars can be grand clashings in the international arena. They can also be fought on a small but ferocious scale, with blizzards of paper as ammunition and cadres of bureaucrats as foot soldiers (*de Villiers 1999, 87*).

This is the first of three chapters about water availability in Canada; here the subject is surface water allocation.[20] Chapter 3 alludes to Canada's apparent richness in surface water, a magnificent endowment of rivers, lakes, snow and glaciers. There are 2 million lakes, crowned by the Great Lakes (which contain 18 percent of the world's water in lakes). The Mackenzie River has the 13th largest drainage basin of world rivers (Canada Natural Resources 2012). But surface water is impermanent; always on the move as it evaporates, dribbles, flows and melts away (Pielou 1998, 1). Climate change, new economic realities, and population increase mean new challenges.

Abundance brings benefits, while transience spawns conflicts and the need for formal methods (doctrines, systems and entitlements) to assure water supplies. Surface water allocation was originally a local solution for individuals who shared water for domestic and economic purposes. Now it is the purview of government to decide who shall or shall not have water rights. Government's legislative and operational activity in this area has a long history and is fundamental to many sustainability issues discussed in later chapters. Along with those directly affected by allocation decisions, people holding 'water rights' (a right to use, not a right of ownership) play the only significant non-government role.

Water Allocation Conflicts

An example gives historical perspective and illustrates both water conflict and legal intricacies (Matsui 2005). The Secwepemc people[21] of the Kamloops and Neskonlith reserves in BC lost their fight for control of irrigation water for agriculture early in the last century. How this occurred shows the "entangled stories of water conflicts among the Secwepemc people, settlers, federal officials, provincial judges, and provincial authorities ... [and] the very complex relations of the four parties, each of which represented different interests" (Matsui 2005, 91). When a provincial official surveyed water rights claims in 1914, he discovered that 300 Native claims had not been recognized. Among these were the Secwepemc's water rights. Their rights supported a significant production of wheat, oats and hay, but they were not informed when their rights were taken away. Even local officials were unaware until water disputes erupted between the Kamloops Band and the Western Canadian Ranching Company in 1906.

In time it was discovered that a delay in filing the Secwepemc's record meant that their official priority for water post-dated a claim by the ranching company. When asked to investigate the conflict, a provincial Board of Investigation found the settler's licence subject to use of water by the Secwepemc people. The Band's right was validated. But the Canadian Ranching Company appealed the decision to the BC Court of Appeal. The court found in favour of the company, citing lack of provincial authority to issue a water licence based on an Indian agent's record. Appeals to the minister of the time proved fruitless and by February 1924 the case was considered closed.

Such issues continue today. In the words of Kanute Loe, an elder with a northeast BC native band who is concerned about the effect of the natural gas industry: "All of a sudden we're having trouble catching fish ... Our rivers are getting harder to navigate ... it's almost like somebody drilled a whole in the bottom of the bathtub" (Hume 2012, S4). Licensing of water rights, then as now, is complex: "elaborate judicial determinations of privileges, claims, duties, and exposures together with governmental rules that create complementary and sometimes conflicting incentives" (Sproule-Jones 2008, 125). Whether existing or future allocation decisions recognize the tenets of sustainability, follow ethical guidelines, and realize the benefits from an integrated, transparent, and adequately informed approach are at issue. To begin this discussion, the next sections explain water use and various Canadian allocation schemes—the basic tenets and new provisions set the stage for future improvement.

How is Water Used?

For most people in the developed world, using water means turning on a tap for household needs. For water managers, uses are withdrawal (including intake, discharge and consumption) or instream uses. Instream uses take advantage of the water where it is; for navigation, migration and spawning of fish, riparian habitat, hydroelectric power generation, recreational pursuits, or simply to allow aesthetic appreciation. Withdrawal uses involve taking water from a source and conveying it to a place for a particular use. Some water may be returned to the source or to the atmosphere through evaporation or transpiration. If untreated, returning water may degrade its source with household or industrial contaminants.

Water may be withdrawn for off-stream uses: manufacturing, agriculture or incorporation into a product such as beer or soft drinks. Water may be transferred to another watershed or be lost in transit from pipes or ditches through leakage or evaporation. Water withdrawals do not always equal total water use or demand as a given volume of water may be 'recycled' or used more than once. All this needs to be taken into account. Today, a "water budget" or "water balance model" tallies inputs and outputs to the system to show how much water can be safely withdrawn. Water use requires a way to allocate water to satisfy many different needs; allocation systems have developed over time, as the next section shows.

Water Allocation Schemes in Canada

For the aboriginal peoples of North America, in earlier times, water was primarily for sustenance use and generally adequate in supply. Settlers brought new uses for water, new conflicts and new concepts of water law (the following description derived from the seminal works of Percy (1988) and Lucas (1990)). Settlers of Eastern Canada introduced the riparian system. American law (influenced by Spanish law) ultimately formed the basis for governance of water use in the West. But Canada has come to have four systems, each locally responsive to broad regions and specific circumstances: the riparian system in Ontario and the Atlantic provinces, augmented by permitting systems; the prior appropriation system in the Western provinces; a civil law approach in Québec, augmented by legislation; and, the most recent, an authority management scheme in the North. Aboriginal water entitlements are important in themselves and as they relate to the various schemes. Also of interest is the concept of apportionment. This chapter provides an overview of the schemes' development and meaning, but does not compare individual provinces and territories.[23]

The Riparian Scheme and Subsequent Permitting Systems in Ontario and the Atlantic Provinces

The riparian scheme comes from the natural right to water of dwellers dependent on a water system—water as a common resource (Shiva 2002, 20). For the early European settlers of Canada, water sources were abundant. Under the riparian scheme, owning land that touches a stream or lake gives, not ownership, but right of access and use of surface water (in defined channels within the watershed)(Figure 3- 1).[24] No deed or title exists. "Reasonable and beneficial use"[25] is permitted for domestic purposes such as drinking, cooking, washing, and supplying an ordinary number of livestock. Certain "public rights" must be respected. Water use for extraordinary purposes (for example, irrigation and the generation of water power) is possible.

Disputes between riparians are resolved by court settlements. For the non-riparian landowner, it is possible to obtain water through a private agreement. After continued water use over a long period, certain rights can be acquired with the consent of riparian owners (Lucas 1990, 6). The concept of riparian rights does provide a simple system, but in time was modified to meet city-dwellers' water needs.

Permit systems developed, mainly to monitor and control major users of water. For example, the *Ontario Water Resources Act* requires a permit when taking of water reaches a certain level or if a use can interfere with a public or private interest in water.[26] The rights and responsibilities of riparian owners continue but become subject to a possible permitting system. Regulations (2004) direct that permits account for natural functions of the ecosystem and water availability, add provisions related to water use and conservation, establish a review system for High Use Watersheds to prohibit withdrawals during summer months, and enable data and reporting (Ontario Environment 2005). Applicants for a permit pay a fee ranging from $750 for straightforward new applications and renewals to $3,000 for applications involving assessment and review of detailed hydrogeological information (Ontario Environment 2007).

The Atlantic Provinces also have or have had riparian schemes. Prince Edward Island relies on groundwater, but the other provinces have both surface water and groundwater sources. While the riparian scheme was dominant in these provinces, now licensing and permitting systems exist. In Newfoundland and Labrador, riparian ownership exists, but is relatively rare (Pratt 1999, 3). Licences do not apply to owners of existing water rights or affect riparian rights of a person requiring water for domestic purposes. Nova Scotia requires an approval for withdrawal or diversion of more than 23,000 litres of water per day, constructing or maintaining a dam, or storing more than 25,000 cubic meters of water. New Brunswick issues permits for surface water withdrawals, including wetlands, but there is no rights-based water allocation.

Figure 3-1: Riparian Doctrine

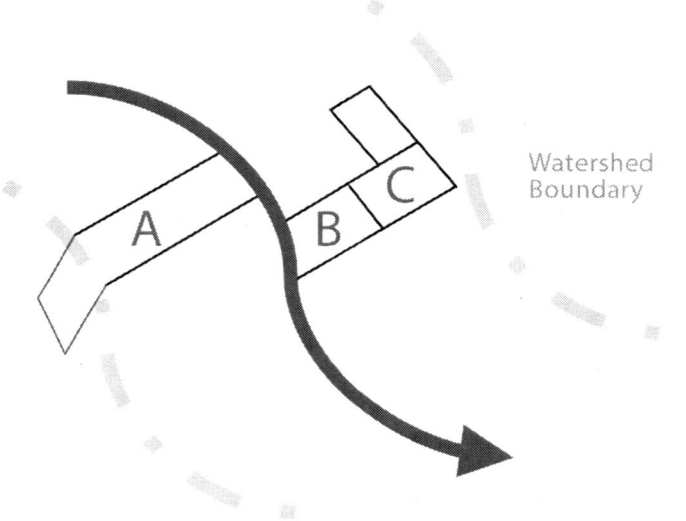

Owner of parcel A can use water only on that part of the land contained within the watershed. If parcel C is subdivided and B sold off then C loses riparian right.

Québec's Civil Law Approach

Québec's surface and groundwater are considered a shared resource, not to be appropriated but a resource that can be used under the provisions of civil law, as opposed to common law. The original Code was enacted in 1866, but a new *Civil Code of Québec* governing private ownership came into force in 1994. Bill 27, *An Act to Affirm the Collective Nature of Water Resources and Provide for Increased Water Resource Protection* came into force on 12 June 2009, bringing with it a number of desirable provisions. Highlights include confirmation of the legal status of both surface water and groundwater as a collective resource, part of the common heritage of Québec. Every person has right of access to safe drinking water. Except under certain conditions, water withdrawal of less than 75,000 litres per day does not require authorization. Approvals are limited to 10 years. The needs of the ecosystem, the population and economic development are to be reconciled when granting permits for withdrawal.

Prior Appropriation Doctrine in British Columbia and the Prairie Provinces

Western Canadian schemes owe much to California's adoption of the earliest prior appropriation doctrine in North America. This section outlines the British Columbia experience in some detail and gives a brief summary relating to the Prairie provinces. While originally subject to the riparian doctrine,[22] the gold rush of the 1850s caused major change in British Columbia water allocation. Increasing demands for water dictated the need for new rules. In particular, gold miners used vast amounts of water: digging in dry areas and then dragging the gravel to the water to be washed or directing water under great pressure against a gold-bearing hillside. Sometimes, they diverted the full flow of a river to mine the riverbeds or built wing-dams to divide the river length-wise. In so doing, the miners staked their claim to water by physically taking—appropriating—the water. The construction of works necessary to divert and take water served as notice to other miners. A miner coming later to a location upstream of another water user would know to allow enough water for those who had staked earlier claims *(Figure 3- 2)*.

Figure 3-2: Prior Appropriation Scheme

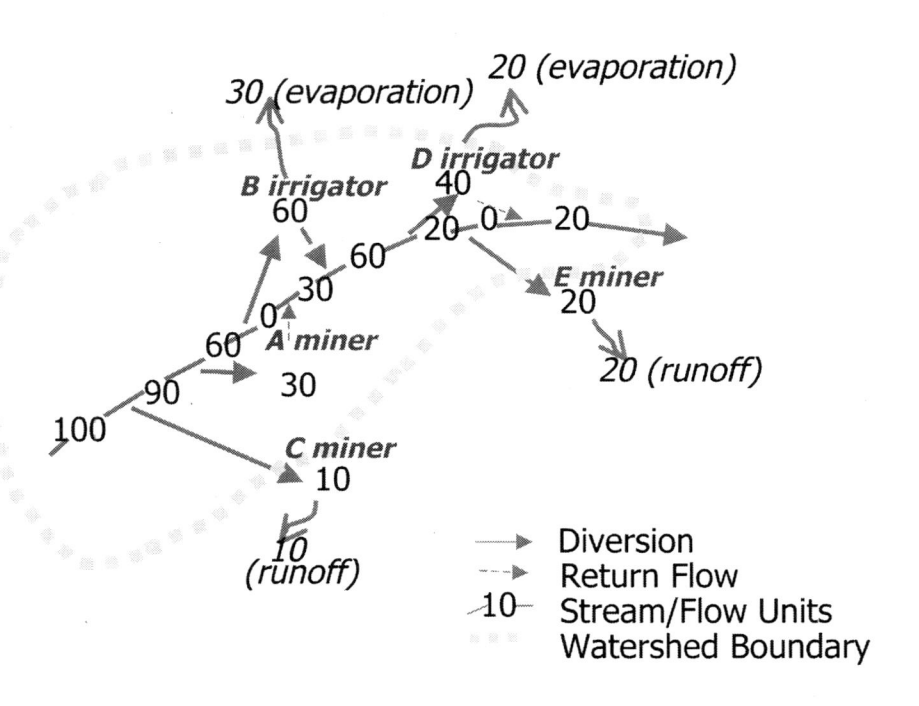

This prior appropriation system uses the "first come, first served" principle, based on occupation rather than ownership. *Qui prior est in tempore, potios est in jure* (first in time is first in right). The system seemed appropriate; private individuals concerned about the security of their rights made decisions far from government control. Like the gold miners, cattle ranchers and fruit and hay growers in BC's Interior required a continuous flow of water to develop their holdings to an economic level.[27]

Governor Douglas reacted to the influx of miners by declaring that all gold mines belonged to British Columbia, following this with a *Proclamation* under the *Gold Fields Act of 1859*. On the basis of a claim, even if not from a riparian owner, rights to exclusive use of defined quantities of water are granted in return for a rental payment to the Crown. This right could be cancelled if no beneficial use is made or if water is wasted. The holder is given the right to sell water, provided a fair and non-discriminatory charge is made.

Subsequent legislation expanded this scheme. For example, the right to the use and flow of all water in any stream is vested in the Crown in the right of the Province of British Columbia (*Water Privileges Act, 1892*); this control had not been clearly specified before and is also in other provincial schemes. A water right is appurtenant (belongs) to the land or mine (the undertaking) for which it is obtained. On sale or transfer of the land or mine, water rights automatically pass to the new owner. This Act confirms the necessity to make reasonable use of water for the purpose of the licence.

In 1909, the first *Water Act* passed after much debate in the Legislature. The Opposition walked out of the House to protest the use of a late night sitting to ram through the bill (Clark 1990, 258)—not the last time proposed water legislation would be controversial. A Board of Investigation subsequently spent ten years hearing claims from persons who believed they held records of water rights and established licence terms. The *Water Act* arguably ends riparian ownership for those who did not file a claim before June 1916 and later amendments reinforce this provision.

In the Prairie provinces, riparian schemes did not serve large-scale water projects well. "Reasonable use" meant shortages for all (Lucas 1990, v). Except in southern Manitoba, the federal government's *North-West Irrigation Act of 1894* rectified this by restricting riparian rights, providing a similar allocation scheme to BC's *Water Privileges Act*, and placing all water rights in the hands of the Crown. The government could then promote agriculture, industry and energy projects. Knafla (2005, 23) tells of Calgary Power who early on gained federal licences for water to build hydroelectric facilities and then offered low electricity prices, a considerable business advantage.

Water rights in southern Manitoba were not regulated until 1930 and also, at that time, the Dominion Government transferred governance over natural resources to the Prairie provinces, the Railway Belt, and the Peace River Block in British Columbia (Percy 1988, 11). In 1938 the *Natural Resources Transfer Agreements* were amended to clarify the inclusion of "water" and "water-powers". All the jurisdictions maintain the right to reserve water for such uses as large-scale irrigation, hydroelectric projects, conservation or water supply. More recently, significant changes have been made to these water rights' schemes.

Revisions to Alberta's legislation in 1999 give a clear process for applicants and guidelines for government managers considering applications. There are a number of notable provisions. New licences issue for a fixed term rather than in perpetuity. Licensees can share their allocation by an agreement and transfer it on a temporary or permanent basis, if approved. The latter provision is not without problems given that Irrigation Districts may then allow their allocation to be used for other than irrigation and agriculture (Christensen and Droitsch 2008, 18-21).

Saskatchewan's water rights scheme followed a somewhat different path. As for the other Prairie Provinces, the *Northwest Irrigation* Act (1894) vested ownership of water with the Crown. The Crown controls water use through a licensing system and does not permit ownership of the bed and shores of water bodies (McDougall 1999). This system became part of the *Water Rights Act*, S.S. 1931. *The Water Corporation Act* (1984) encouraged a more integrated approach to water management. With this new act, while pre-1984 rights remain intact, much was left to the discretion of the Water Corporation when dealing with water allocation; the Act did not establish which uses should have priority.

The "first in time, first in right" rules did not have to be applied if the minister deemed it necessary to reserve water or was willing to pay compensation where cancellation was in the public interest (Percy 1988, 39-42). In 2002 the Corporation's responsibilities transferred to a new Saskatchewan Watershed Authority and in 2005, the *Saskatchewan Watershed Authority Act* was adopted. Riparian landowners retain the right to use water but must seek approval for diversion works. Allocation provisions do not appear significantly changed in this Act. Other provisions, spurred in part by the North Battleford situation (see Chapter 7), include: watershed planning, prevention from contamination of surface water bodies, and protection of aquifers, the health of aquatic and riparian ecosystems, and of drinking water sources.

Manitoba's *Water Rights Act* vests ownership of water and all rights to divert and use water in the Crown and has similar allocation provisions to other acts described. The use of water for domestic purposes is exempt from licensing. The 2005 *Water Protection Act* adds provisions relating to protection of the aquatic ecosystem and preparation of water management plans. A Water Advisory Council, formed in 2007, will co-ordinate and oversee the work of all provincial advisory bodies on water protection and give advice to government (through its Water Stewardship Department), stakeholders and the public on water-related policies, programs, plans, and legislation.

All of the prior appropriation schemes described come with responsibilities in the form of water licence terms and conditions.[28] To summarize the discussion up to now, riparian and prior allocation schemes of water allocation differ as follows (Ontario Environment 2005; Percy 1988, 18; Scott 1991, 348-351; Thompson 1999):

1. **How is a "water right" derived?:** A riparian's right to water derives from and is tied to ownership of riparian land and is not affected by Crown ownership of the banks and bed of a stream; under prior appropriation, someone who diverts the water and applies it to a beneficial use acquires the right and is given a licence. In permitting schemes, such as Ontario's, certain takings require permits.

2. **Can you lose a water right?** Riparians never lose water rights so long as they own riparian land (unless a statute removes the right); non-use can destroy the right under prior appropriation (in permitting schemes such as Ontario's, where taking of water by any person interferes with any public or private interest, a person may be prohibited from taking water without a permit).

3. **Where can water be used?** Riparians use water on adjacent land within the watershed; appropriators and those under permit may use the water anywhere, but the right is appurtenant to a specific parcel. If a watercourse changes direction, the riparian loses their right to water while under the prior appropriation scheme, they do not.

4. **How much water can you use?** Riparians share in the use of water with no fixed quantity assured them over time. They must return the water diverted back to the watercourse undiminished in quantity and quality. Appropriators with licences and persons under permit obtain the right to a specific flow or quantity in perpetuity or for a limited time based on the scheme.

5. **What happens if there is not enough water or conflict between users?** Riparians suffer equally in times of shortage but have recourse to the courts if conflict arises; appropriators suffer unequally according to seniority and permit holders are subject to specific conditions based on administrative procedures and may resolve conflicts through dispute resolution or plans.

6. **Is there any flexibility in use?** Riparians have more flexibility for "reasonable" use; appropriators or permittees must use water for a specific use and can see their use affected by water plans or conservation guidance.

7. **Can you transfer your water right to another property owner?** In a riparian scheme, transfer of water rights is usually limited to persons owning riparian land. In an appropriation or permit scheme, transfer

of rights passes with the land title no matter where the land is (unless otherwise established or prohibited by legislation).

8. **Are you guaranteed a certain quality of water?** Riparians can demand water of a certain quality while with the prior appropriation and permitting schemes, there is usually no guarantee of quality.

Northern Authority Management Scheme

Northern water law is the most recent allocation scheme and exists notwithstanding the effect of future land claims settlements (Percy 1988, 48-72). Riparian rights are limited through the reservation of strips of land, 100 feet in width measured inland from the high water mark, under the 1985 *Territorial Lands Act*. The 1970 *Northern Inland Waters Act* gave most legislative authority to the federal Department of Indian Affairs and Northern Development. The Act vests administrative responsibility in Territorial Water Boards. Provisions of interest include:

- The Crown may reserve water in any area and choose not to grant any licences.
- Allocation is according to a priority of use (as established by the Water Board), rather than by date (prior allocation scheme); the date of application establishes precedence.
- Officials consider both water quantity and water quality.
- A licence applicant must prove that their proposed use will not adversely affect either an existing licensee or another applicant having higher priority uses. The applicant must also demonstrate that licensees who have a lower priority use or who are adversely affected will be compensated.
- Any water use or diversion requires a licence; licences may be granted for a period not exceeding 25 years, subject to the appropriate conditions.
- The Water Board retains the authority to amend the licence should water shortage become a problem, water quality standards change, or other public interest issues arise.
- Transfer of water rights requires transfer of the undertaking to which the licence is appurtenant.
- Public hearings are mandatory in the consideration of water licences

Apportionment

Another concept relating to allocation schemes is apportionment, the negotiated division of transboundary river flows between or among jurisdictions (Quinn et al. 2003, 2). Apportionment applies to transboundary flows in dry areas (for example, the Prairie Provinces Water Board 1969 Master Agreement on Apportionment) and usually occurs on a 50/50 basis between upstream and downstream jurisdictions. Possible future challenges for apportionment agreements include: water quality problems and vested interests relating to water use opposing water conservation measures (Brooymans 2011, 108), application to the MacKenzie River basin, the effect of aboriginal title to water previously ignored, and the enforceability of the interprovincial agreement for eastward-flowing Prairie rivers (Quinn 2003, 4).

Aboriginal Water Entitlements[29]

Aboriginal title is derived at common law from the historic occupation and possession of tribal lands. Based on western law, rights to water became an integral part of aboriginal title, but these rights may be limited to historic and traditional uses (Bartlett 1988, 7, 8). These rights may include rights to travel, navigation, use of water for domestic or harvesting purposes or "for spiritual, ceremonial, cultural or recreational purposes" (Laidlaw and Passelac-Ross 2010). Aboriginal water rights depend on interpretation of riparian rights, treaties and agreements (and their variable adoption across Canada), as well as their wording, and whether "waters" and "watercourses" were specifically surrendered. Aboriginal peoples of the time may not have understood that the treaties meant surrender of water.

Subsequent case law, often relying on American jurisprudence, has developed current understanding. For example, case law suggests that water rights on a reserve are to be construed with regard to the intent for which the reserve is set up (Bartlett 1988, 10, 39). More recently, settlement of land claims gives clarification to some situations; for example, the Nisga'a Treaty provides that ownership and regulatory authority over water be retained by the province and continues existing water licences (Nowlan 2004, 21-22).

But western law contrasts with indigenous law that sees water as the giver of life, a sacred trust—not to be owned, but to be shared with no one interest paramount and with the obligation of protection for present and future generations (Innes 2010). Kate Kempton practices Aboriginal rights and environmental law, representing First Nations in negotiations and civil litigation. In discussing how the Great Lakes Annex Regime could affect aboriginal and treaty rights in Ontario, she reviews rights to water and dependent on water. Her analysis takes the reader from an explanation of inherent indigenous rights through an analysis of the patterns of treatment of aboriginal rights to suggestions for a hoped-for future. Inherent indigenous rights derive "from existence (being here) and custom (adaptation of a way of life to perpetuate existence or survival as peoples). Custom (or customary law) is in turn derived from the relationship with the Creator and the understanding of why and for what purposes the Creator put a people here (in their own place in the universe)" (Kempton 2005, 13).

Case law on whether reserve lands include riparian rights is considered inconclusive (McNeil 1999, 6). Whether provincial water law applies on reserve lands is the subject of divergent opinions (Bartlett 1988, 127; Tyler 1999, 12) and in reality, even where licenses exist, they are not always honoured (Walkem 2007, 305). This uncertainty has a long history as illustrated at the beginning of this chapter. The future may see self-determination and control over development and restoration/maintenance of cultural identity (Innes 2010).

Challenges to the Allocation Schemes

The challenge in water allocation is to operate within historic schemes (and their consequent decisions) in a world far different from the time when water was plentiful and sharing water a local matter. Water "rights" now created go well beyond the intent of the original common law concepts. What do we need to do to improve? Riparian and prior allocation schemes have their drawbacks, as used historically. Riparian schemes, while acknowledging water rights in a simple, but not particularly flexible, manner are inadequate to the stresses of economic development or population increases. They engender uncertainty as, in times of drought, and now of climate change, each riparian has to share the consequences equally with others using the water source. This system does encourage informal arrangements for rights-holders to work together, promoting a common good ethic, but does not lend itself well to integrated management efforts. Requiring that water be returned to downstream users without degradation (a tenet of the riparian doctrine and now seen in the Québec scheme) promotes ecosystem integrity (even where there is no specific legal protection for instream rights) and prevention of pollution. All these factors lead to the need for other schemes, such as permitting, in which a better response to the problems mentioned is possible.

A prior allocation system provides an orderly system for recording and management of water rights.[30] Attention to the nature, seasonality, return intervals and various uses eventually translates into protection of activities dependent on water. During water shortage, it clearly delineates priority, whether the user is above or below another's diversion point (although it does not distinguish between uses in these circumstances). Historically, the system did not encourage water conservation for instream, intrinsic, recreational values or water demand management. In fact, it could encourage water waste and result in over allocation of a single water source where monitoring was not adequate.

This legacy creates problems for allocation today, despite recent legislative changes that direct consideration of instream uses. Response to drought and climate change is possible, but those who came last will suffer most. The Northern Authority Management scheme has the potential to avoid the perils of the "first come, first served" regime, but is challenged to determine which water uses should receive highest priority. This scheme does benefit from its consideration of water quality and quantity together, from fixed licence terms that allow for future review, and from the requirement for a mandatory public hearing.

Is there a better way? Challenges for the future include creating allocation systems that will recognize the need for water conservation, consistently consider relevant water quality, and determine the effect of groundwater's connections to surface water. Second, greater monitoring and enforcement are essential to ensure that licences are acquired, or where existing, are in compliance (Kreutzwiser et al 2004). Third, there is much work to be done to fully recognize aboriginal water rights (Nowlan 2004, 37). Finally, greater public and stakeholder input in licensing and permitting activities would reveal historical context and public concerns.

Planning approaches (see Chapter 11) show promise in the resolution of issues relating to water allocation. They allow consideration of water quantity and quality together as well as instream and withdrawal uses together. Collection and mapping of information highlights threats to the resource and makes this available to all those involved. Water allocation planning fosters better water demand management and sets the stage for water licensing at a local level. It benefits from greater public awareness, engagement, and commitment by all to a more clearly defined future. In the words of the National Round Table (ceased operation March 31, 2013): "The emergence of collaborative governance models provides a potential opportunity to improve the way we manage water in Canada and brings the flexibility required for addressing regional and local particularities. It will require, however, strong leadership from governments to create the conditions necessary for these collaborative arrangements to be successful" (National Round Table 2010, 56).

Conclusion

This chapter introduces Canadian surface water allocation practices and asks how they measure up today. Such discussions sometimes seem less interesting as they must cling to their technical and legal underpinnings. But they are essential as the basis for other issues.

The schemes work reasonably well for their original purpose: allocating water for various uses. Often this is the result of vigilant and creative intervention by water managers and elegant compromise on the part of those affected. Societal objectives of water law, according to Lucas (1990, 14), should be to maximize the value of the resource to society; protect and promote public water uses; clearly order private rights; and ensure fairness, flexibility, and efficiency in water allocation and management. These objectives are echoed in this book's call for sustainability, integrative and collaborative decision-making, and an ethic of care for water. Sustainability, in the context of this chapter, would be typified by protection of the resource to ensure renewability and quality (including interconnections between surface and groundwater), ability to adjust to drought and climate change, recognition of instream uses, and attention to aboriginal entitlements. While recent improvements to legislation go some way to enabling these goals, there is much to do at an operational level, particularly given historic allocations. Decision-making through water allocation planning and transparent allocation processes are means to this end and give opportunities for local involvement in a primarily government function.

CHAPTER 4: PROTECTING GROUNDWATER: A COMMUNITY'S STORY

> Drawing on my own well added something to my sense of the house as an ark and a refuge in time of trouble. It became a rite to drink a glass from it whenever I had been away and come back. ... Of course no well is inexhaustible. I showed my city-bred ignorance the first summer by setting up a whirling lawn hose and letting it run all day, and then being much dismayed when the water ran out. ... Water, like food, is essential. The lack of it breaks down one's faith in life itself (*Sarton 1968, 141-144*).

Sarton's well brings her a short-lived sense of security and greater awareness of a precious resource. In Canada, this resource is now under threat from urbanization, climate change, energy production, agricultural developments, and contamination sources (Council of Canadian Academies 2009, 183). This chapter briefly describes groundwater's characteristics, use and the government role in its protection. Then, one community's long and complicated path to groundwater protection illustrates the importance of "social learning" and collaboration as people work together to protect their groundwater. Provincial and local agendas merge and conflict displaying the "politics" of water management.

Groundwater Defined

For most of us, the dominant characteristic of groundwater is that its source is invisible and hence, somewhat mysterious. Out of 400 citations in 100 Canadian short stories and novels, 57 percent of the images of groundwater are negative; only 22 percent are positive (Kohut 1999). The others are neutral or contain both negative and positive images. These fictive sources display dark images, mysterious circumstances, uncertainty and fear. Al Kohut, then Head of the Groundwater Section of BC Environment, says: "Wells may be places of negative feelings and uncertainty, death and decay, a tomb, a place of cold, darkness, silence and mystery" (Kohut 1999, 433). He wonders if the pervasive nature of such images could compromise our ability to manage groundwater. The reality is a resource that does provide complex problems and, indeed, a resource that requires some explanation.

How does groundwater form? About 10 percent of rain or snow percolates through soil reaching a zone of saturation; the upper surface of this zone is the water table. The water that saturates the ground, filling all available spaces, is groundwater.[31] It normally moves very slowly (30 metres per year typically), usually from higher to lower elevations, and is affected by the material it travels through. This material may be grain particles of rock or soil, or crevices or fractures in rock. When enough water accumulates to be "useful", when "water is abundant enough, and can flow through the ground fast enough," an aquifer forms, usually within 100 metres of the surface of the earth (Pielou 1998, 4). Aquifers may be confined, held by materials like sandstone or limestone, and under greater than atmospheric pressure. The pressure sometimes causes a flow to the surface (artesian well). Aquifers may also be unconfined, made up of sand and gravel, saturated with water up to the level of the water table. The levels of these aquifers rise and fall with discharge and recharge and based on barometric pressure.

Aquifers may be only a few hectares in area, or may underlie thousands of square kilometres of the earth's surface. In a few days or in many thousands of years, groundwater will discharge, reappearing on the surface and becoming part of rivers, lakes, or streams (Canada Environment 2010c). Springs (natural groundwater discharges) are particularly important sources of water for communities, fish hatcheries, commercial water bottlers, irrigation, and watering livestock. Of course, climate significantly affects how much water goes into an aquifer. Groundwater quality depends on the kind of material that holds the water and the pollution sources that affect it.

Groundwater Use

Groundwater use is variable and sometimes unsustainable. Windmills, used on the Prairie provinces for over a hundred years, extract groundwater very slowly. Centrifugal pumps increase this rate, but they are expensive and only operate if the water is within 20 metres of the water table. Today, cheap electric power pumps groundwater, drawing from 10 to 2000 litres per minute, sometimes more water than can be replenished.

Groundwater has many benefits. It may be the main or only source of water for both domestic use and irrigation. It performs an important ecological function, maintaining streamflow during dry months. Groundwater is often available at the point of need for less than the cost of constructing pipelines to bring water from elsewhere. Where protected from pollution sources, it is usually free of suspended sediments, pollution, and bacteria-causing disease. It is particularly useful in dry areas. The *Atlas of Canada* shows that aquifers composed of thick deposits of sand and gravel supply water for large municipalities such as the Kitchener-Waterloo region in Ontario, the Fredericton area in New Brunswick, and municipalities in BC's lower Fraser Valley. The sandstone aquifers of Prince Edward Island supply that entire island's water. Even though there may be greater groundwater supplies than surface water, as of 1996 groundwater served only about 30 percent of the Canadian population for domestic use (Canada Environment 2004a). Total groundwater use in Canada is 2 percent of the fresh water withdrawn (Commission for Environmental Cooperation 2002, 2-3).

Manufacturing, mining, thermal power generation, aquaculture, and increasingly the water bottling industry (see Chapter 10) are the biggest consumers of groundwater (Côté 2006, 3). For example, groundwater from the fractured rocks of both the Winnipeg and Montréal aquifers supplies water for industry. There are many small aquifers serving farmland and rural homes. About 90 percent of rural residences get their water from wells, and well water is used for livestock. Irrigation from groundwater, however, varies from 48 percent in Manitoba to one percent in Alberta (Canada Agriculture and Agri-Food 2003).

Unfortunately, there are also problems. Some aquifers studied by the Geological Survey of Canada show stress (Rivera 2005, 8). For example, Manitoba prohibited new groundwater permits in 6 of 13 sub-basins in the Assiniboine Delta Aquifer (Nowlan 2005, 47). Falling groundwater levels in the Kitchener-Waterloo area are a major concern for about 250,000 people relying on this source (Commission for Environmental Cooperation 2002, 4). Contaminated aquifers are difficult and very costly, if possible, to clean up. Aquifers can also contain high concentrations of naturally occurring dissolved minerals (Pielou 1998, 47). Other problems may include: conflicts where aquifers cross jurisdictional boundaries (Abbotsford Aquifer), land subsidence where significant withdrawal of groundwater causes collapse of air-filled former aquifers (currently evident in Mexico, the U.S., Japan, China and Thailand), and salt water intrusion caused by over-use of groundwater near seacoasts.

Canada's knowledge about groundwater is growing, but is still far from sufficient. *The Canadian Framework for Collaboration on Groundwater* (Rivera et al. 2003) proposes a national groundwater inventory including assessment of "key" regional aquifers.

There is no overall assessment of groundwater quality, although there are local and provincial assessments. Reports suggest concern relating to nitrate levels, bacterial contamination in rural wells and naturally occurring trace minerals such as arsenic and fluoride (Van Veen et al. 2003).

Groundwater Governance and Management

Questions arise as to whether governments actually care about groundwater. The Federal government has jurisdiction over aquifers that cross interprovincial or international boundaries; otherwise jurisdiction is in the hands of provincial governments. Federal initiatives have been limited: proposals for management in the 1987 *Federal Water Policy*, much later the aforementioned *Canadian Framework* (Rivera 2003) and the Canadian Council Ministers of the Environment (2010a) *Review and Assessment of Canadian Groundwater Resources, Management, Current Research Mechanisms and Priorities*. This apparent disinterest may occur because groundwater supplies fewer domestic users than fresh water and is less familiar to urban than rural dwellers. And groundwater poses the same administrative problems as surface water for the federal government; few of its issues fit within the mandate of a single department (Canadian Geoscience Council 1993). More recently, the Geological Survey of Canada received $4.4 million a year from 2005 to 2008 to map and study Canada's main aquifers. Groundwater assessments are now complete for 9 of 30 key national aquifers (by 2009).

If you review legislation, you find that groundwater is sometimes referred to as "water" or "environment" (Porter c1999, 4). It is often considered separately from surface water, ignoring the linkage between the two, and threatening sustainable use. In part, this reflects the notion of "absolute capture"; historically landowners could extract groundwater from under their land with no concerns about how this affects neighbouring wells. To counteract this, groundwater is usually vested in the Crown in the right of a province; in other words, legislation states that a province owns and has always owned the groundwater. Provincial laws and regulations may then restrict its use and specify protection measures. Groundwater management exists in a number of places throughout Canada. The main tools are inventory and monitoring, allocation mechanisms such as permits and licenses, drinking water quality and waste management standards, planning measures, and programs for education and awareness.

Other protection measures include: environmental assessments, local government plans that restrict development over groundwater recharge areas, zoning bylaws that control land uses and the amount of impervious surfaces, and stormwater management guidelines. For example, Prince Edward Island's total dependence on groundwater has meant early adoption of strict controls relating to the replacement of older oil tanks (described in Chapter 6). There are many different protection schemes across Canada. The schemes are not comprehensive and, in the case of BC, have taken a long time to develop. To better understand groundwater, its protection, and governance, we turn to one area in greater detail.

BC's Islands Trust Area and Groundwater Dependence

The Islands Trust Area covers the Gulf Islands and waters between the mainland of BC and southern Vancouver Island, as far north as Comox. These islands provide an excellent example of groundwater issues. While there are some lakes and streams, groundwater is the primary water supply source.[32] Faulting has affected most of the region; groundwater moves and is stored within the bedrock's open fracture zones. These fractures (observed on air photos) are widely distributed and appear interconnected, possibly suggesting continuity between groundwater sources on a regional scale. However, major faults can also act as a barrier, short circuiting groundwater systems. Bedrock wells supply groundwater for domestic use, community wells and small farming operations. Well depths range from 30 to 150 meters; drilled by the air rotary method, the cuttings are removed from a borehole using high pressure created by an air compressor at the surface. Well yields normally vary from 10 to 20 litres per minute. Groundwater supplies are replenished through recharge areas that require protection, particularly upland areas where other resource uses may generate pollutants.

The islands' groundwater sources are not without problems. Most wells are unlined, interacting at deep and shallow levels, so causing quantity and quality problems. Quantity problems may not be apparent until wells run dry in the summer. Quality varies considerably; the water may be too saline, have unacceptable arsenic and fecal coliform levels, or contain hydrogen sulfide gas (odour and corroding pipes). Poor construction practices have included inadequate wellhead protection and well disinfection, improper well sealing, or failing to undertake necessary sampling and analysis of well water. Where problems exist, people may use the available water for washing and bathing, and then import drinking water. Given the number of issues, groundwater's management and protection is crucial. It is equally important to understand barriers to adoption of necessary measures. A case study develops this discussion.

Protecting Groundwater on Hornby Island

Until recently, British Columbia was the only jurisdiction in North America that did not have some form of groundwater regulation. This case study tells the story of what happened when the Province decided to test its proposals for groundwater governance by undertaking a pilot project on Hornby Island. The story has many twists and turns, covers a period of more than 16 years, and results in solutions that might not have been expected by everyone involved. It began when the Province went to the public with a Discussion Paper *Stewardship of the Water of British Columbia* (1993-1), suggesting that a successful scheme for groundwater management should be comprehensive, involve both voluntary and regulatory actions, and be site-specific. The paper suggests certain actions (vesting of groundwater in the province, creation of groundwater management areas, province-wide standards, water quality provisions and transboundary agreements); would these actions work and would they be acceptable in every community? Hornby Island was one of two pilot project areas chosen (the other being Pender Island).

Hornby Island is the smallest (2,990 hectares) of the islands to form an Islands Trust Local Government Area and is relatively isolated; you need to travel on two ferries and drive across Denman Island to reach it. Its residents (1,076 by the 2006 census, likely quadrupled in the summer time) are passionate about protecting this island which is often described as the most beautiful of the Gulf Islands. Whether people come for a family summer, retirement, to escape, to visit, or to start a new life, Hornby Island is a special place. Land use planning and the regulation of development on the island is the responsibility of the Hornby Island Local Trust Committee, one of 13 independent local governments that make up the Islands Trust. This federation's mandate is to "preserve and protect the Trust Area and its unique amenities and environment for the benefit of the residents of the Trust Area and of the Province generally, in cooperation with municipalities, regional districts, improvement districts, other persons and organizations, and the government of the Province." The challenges portrayed by that mandate are shown by what follows.

To start the Hornby Island Pilot Project, the Ministry of Environment, Lands and Parks (Environment) prepared a summary of available information about groundwater (Hornby Island Pilot Project Committee 1994). This report shows how in the central area of the island, two large parcels of public land, the regional park on top of Mount Geoffrey, and vacant Crown land to the east of the park, are the main sources of recharge for the groundwater supply. Rainfall and snowmelt replenish groundwater during the fall and winter in these and other water catchment areas. Groundwater is mainly in open fractures or breaks that occur within the sedimentary rocks which underlie the whole island. These fracture openings occupy less than one percent of the rock formations, a very small area in which to store and transmit groundwater; hence, supply problems develop. Interconnection between the fractures allows groundwater to travel from one part of the island to another, sometimes spreading water quality problems.

Existing wells drilled in the bedrock are generally 15 centimetres in diameter and range in depth from 15 to 75 metres; the deepest well on record at the time of the pilot project was 152 metres. Individual well yields depend on the location, number and size of water-bearing fractures encountered during drilling and the position of the water level in the well. Most yield less than 0.3 litres/second. Groundwater is also found in thin unconsolidated deposits of sand and gravel overlying the bedrock areas along some creeks. In these cases, shallow, large-diameter wells produce groundwater from surface springs.

Hornby Island's groundwater was found to have both quantity and quality problems (Hornby Island Pilot Project Committee 1994).[33] These problems explain why Hornby Island was chosen from a number of interested communities as a pilot project to test provincial proposals. Certain other factors were also important: groundwater can be studied in isolation in an island setting, is virtually the only source of water for the island residents and visitors, and can be seriously affected by major developments. Any management scheme could be replicated on an island-by-island basis. Volunteer assistance would be available; some Hornby Island residents already had considerable expertise. Finally, the project would support the mandate of the Islands Trust "to preserve and protect ... the environment and ... amenities of the Trust area." In response to the announcement by the Minister of (Environment) on July 23, 1993, the Chair of the Islands Trust welcomed the initiative and the first step forward in groundwater management.

Besides testing the effectiveness and practicality of government proposals, project goals were to encourage a local government role in groundwater management, consult stakeholders having a direct interest in groundwater, inform those preparing draft legislation, and consider the principle of integrating provincial and local land management. Project work was led by the Hornby Island Pilot Project Committee including government staff (Environment, Health and the Islands Trust), two elected trustees, and one member of the public. Local well-drillers chose not to participate. The project was to extend from September to December 1993 to inform new legislation proposed for 1994.

The Committee held an exploratory meeting, an on-island public meeting, and two evening workshops. At the public meeting on September 15th, 1993, government representatives made presentations to about 60 interested residents and listened to their concerns. They outlined the various aspects of the project: the need for groundwater information, health concerns regarding water quality, and the *Stewardship* proposals. The history of surface water management in BC and the current allocation scheme were explained.

At a public workshop one month later, 30 participants met in small groups to discuss issues and bring forward proposals. Each group examined one issue (such as protecting the watershed or integration of surface and groundwater management), defined the problems, and then gathered comments for their report. The meeting provided "imaginative and innovative" assessments of the situation (BC (Environment) 1993). Following the meeting, recommendations were summarized and distributed to meeting participants. There are many lessons about groundwater management sensitive to a local situation; community concerns were not always the same as the government's province-wide areas of interest.

First, meeting participants requested designation and protection of the groundwater recharge and discharge areas, particularly the Crown land parcel on the Island's Mt. Geoffrey, and monitoring of squatters' use of this land. They wanted no controls over drainage activities on private lands, but actions to remove abandoned vehicles and prohibit other goods hazardous to water quality. They also asked for monitoring of air quality for acid rain. To empower the community, a Water Board should manage and protect the groundwater resource. In an approach common to islands in the Trust area, the Hornby Islanders wanted to involve the "whole" community, including visitors, through an active education program with workshops, newspaper articles, and community-gathered data collection. By doing this, they hoped to encourage conservation by rainwater storage, individual cisterns, capping of artesian wells, regulation of excessive use of water, improvement and maintenance of individual waste systems, and consideration of a community reservoir. Restoration of contaminated aquifers by re-planting de-forested areas and limiting well sites near saltwater was also requested.

Licensing and regulation were to be limited to large capacity wells. Well records should be kept and there should be a small charge for registration. If annual costs for registration and licensing were above $20, then an annual test of water quality should be included. Metering of residential wells was seen as unnecessary. Finally, the meeting participants looked to the Hornby Island Local Trust Committee to regulate new development according to water availability and to control the number of visitors to the island. Taken together, these views emphasize comprehensive groundwater protection tailored to the local level: land use management, local involvement, education and information provision, some form of regulation for specific problems, partnerships between government and non-government participants, and little or no fee payment. Following the meeting, the recharge area on Crown Land was protected by a Reserve (or "Notation of Interest"). Future use of this land would take into account the importance of groundwater protection. Gates were to be erected at access points to control illegal camping.

The Hornby Island Pilot Project Committee considered the input of the islanders and proposals of the *Stewardship of the Water of British Columbia* Discussion Paper and drew a number of conclusions. Groundwater should be vested in the

Province of BC as it is a public resource, but the established uses of groundwater must not be taken away. A groundwater management area should be designated and a groundwater management plan prepared in identified problem areas. The plan should be reviewed by the community before acceptance by the Minister; implementation of certain measures should be delegated to the community. Regulatory measures should only be used where there is a recognized need (such as stress on quantity of groundwater or diminishing quality), and then selected tools should be used; for example, regulation of wells exceeding extraction of 7000 litres per day.

Regulatory measures should also include permits for new wells, servicing of existing wells, or abandonment of wells; control of land use activities generating pollution near well-heads; standards for well-head construction; and refusal of use for drinking water purposes from contaminated aquifers, unless treated. The Committee agreed that the Notation of Interest on Crown land should be maintained to protect the recharge area. A community watershed designation would allow control of pesticides and chemical fertilizer use, storage of car bodies and other hazardous materials, inappropriate land drainage, sewage disposal and clear-cutting of trees. Certification of well drillers or licensing (except as above) should not be introduced and no fees should be charged. Finally, emphasis should be placed on the provision of information and education about groundwater.

The Committee's recommendations respond to the provincial government's more regulatory approach. Nevertheless, they reflect many of the community's concerns. Unfortunately, the Province of British Columbia chose not to go forward with groundwater legislation at that time given concerns elsewhere about charging for groundwater use. Despite this, Hornby Island residents and their local government, sometimes assisted by provincial government staff, were not idle. From 1996 to 1999, over 800 surface, marine and well water samples were taken by volunteers (showing that 20 percent of 167 wells had unacceptable fecal coliform counts) and a university student project identified groundwater pollution from older and inadequate wastewater treatment systems. A group of Hornby Island residents identified necessary steps for a management plan and provided recommendations for the island's Official Community Plan, then under preparation. Initial objectives for a Groundwater Protection Strategy included information gathering, development of education tools, and facilitating establishment of a performance assurance system for wastewater disposal systems.

In 2001 the Ministry of Water, Land and Air Protection (former (Environment)) released a study of the Hornby Island aquifer as part of its province-wide aquifer classification project. It found that the entire island is vulnerable to contamination, being one of two areas of the province of highest priority for groundwater protection (out of the 438 aquifers studied at that point). Between 2000 and 2002, activity increased. A Hornby Island Advisory Groundwater Protection Committee formed and prepared a Groundwater Protection Strategy. The Committee included various groundwater-related groups facilitated by the Islands Trust and the Ministry of Water, Land and Air Protection. Evidently there were sensitivities to overcome: adoption of the term "advisory" in the Committee's title and concern about the term "management"; how the new committee would fit in due to the number of different initiatives in the past; and ensuring that this was an 'island' committee, representing the island to the Islands Trust and other agencies, not the other way around (Kneffel 2002, 2). Particularly notable among the on-island activities was work of the Water Stewardship Project, sponsored by the Heron Rocks Friendship Centre Society, which monitored creeks, ditches and beaches for fecal contamination levels, lobbied for necessary changes, educated the public, and "paved the way steadily building trust in the community in what can sometimes be a delicate issue" (Kneffel, iii).

The Advisory Groundwater Protection Committee successfully acquired funding. By providing a central access point, they were viewed more favourably by government, receiving a grant of $26,500. Before the Committee began work, three other important issues were resolved: the committee and off-island agencies agreed that confidentiality would be maintained regarding any data collected; no action would be taken against an owner of an inadequate water system; and the expected contribution of each off-island agency was defined.

A hydrogeological study confirmed that much of the island is a recharge area, particularly areas of high elevation, and that groundwater originates on the island (Allen and Matsuo 2002).[34] Fractured bedrock, sandstone, mudstone and conglomerate facilitate potential transport of contaminated water. There is saltwater intrusion in one area, but no significant toxic metals contamination in the areas tested. The study recommended land use planning and management of development density. These results were presented to a meeting of about 90 people, as part of the 'Literary Lunch' series. Other public events included a meeting and several 'Co-op Porch' educational events, the latter held on the Co-op Community Store deck and featuring information brochures and a painted display courtesy of the Hornby Island Water Stewardship Project.

In the end, one initiative (at a local government level) provides some certainty for protection—the Hornby Island Official Community Plan (Hornby Island Local Trust Committee 2002). The plan seeks protection of fresh water and groundwater recharge areas and the Mt. Geoffrey area as a groundwater protection-ecosystem management area. It recommends that a Management Plan for these areas be prepared and that development be limited where the groundwater geochemistry study identifies problems. Other measures include: protection areas for community well-sites and a community-based resource protection/management committee to co-ordinate and implement a Groundwater Protection Strategy. Large-scale desalination plants are prohibited. The plan encourages alternate methods of water collection: rain water catchment, lined dugouts, and well reservoirs for uses such as filling swimming pools. The plan requests the provincial government to consider designating Hornby Island as a Groundwater Management Area, with provision for local participation.

A Groundwater Protection Forum (environment interest groups, local and regional government, the ratepayers' association, a residents' group, other interested community members and companies) was created to implement Phase III of the project. This group assisted in completion of the *Hornby Island Groundwater Protection Strategy*, the result of a number of iterations and intended to be a living document (Kneffel 2002, 14). Its quantity-related proposals are protection of wetlands on public and private property, preservation and enhancement of groundwater recharge through awareness of land use practices such as clearing and ditching, and increasing water conservation awareness and practice. Quality-related proposals are promotion of proper wastewater treatment, correction of failing septic systems, and preventing contaminated water entering wells or around well casings. Aquifer contamination resulting from abandoned or poorly constructed wells, household and land use activities, and inappropriate agricultural practices is to be addressed.

The Strategy also recommends development of an implementation program, collection of further information, and hiring of a paid Coordinator. British Columbia has subsequently passed groundwater protection regulations and intends to regulate groundwater use in priority areas and where there are large groundwater withdrawals (BC Environment 2008, 49). In response to proposals by the province in 2010 (legislation proposed for 2014), the Hornby Island Local Trust Committee supports regulation of large withdrawals where problems are demonstrated, recommends local involvement in plan preparation, allocation sensitive to the ecosystem, drinking water and food production needs. It further indicates willingness for designation as a plan area and participation in a pilot project (Hornby Island Local Trust Committee 2010).

The attempts of Hornby islanders and others to protect groundwater suggest a number of lessons. While these come from a small and isolated community, they provide both guidance and hope to those who seek better groundwater protection. The lengthy consultation period brought forward community concerns and acceptable solutions. A provincial strategy might not uncover: concern about squatters, abandoned vehicles, and other goods hazardous to water quality on the Crown land recharge area; the desire for local leadership; and a call for reliance on education and conservation measures. Limitations on population growth and visitors were contemplated. Particularly important was the issue of confidentiality of data collected and guaranteed freedom from possible action, where problems were uncovered. As the project progressed, it became evident that local government needs to play a significant role, given the impact of land use activities on groundwater quality and quantity.

There are also a number of problems and risks with the approach and time period taken. Real action has taken and will require many years to occur. Problems that were not serious at the beginning may now be more troubling or more frequent. This approach would not be suitable in an area requiring urgent action or if an aquifer crossed jurisdictional boundaries. Finally, the project struggled, but ultimately found success, in dealing with the number of different initiatives and interest groups involved.

Learning from the Hornby Island Experience

Water management usually deals with clearly defined problems and proposes technological or institutional fixes (Pahl-Wostl 2002, 394)—on Hornby Island, the provincial government proposed the latter. The environment–technology–human systems were approached along separate paths, characteristic of most environmental management in the past (Pahl-Wostl 2002, 395). By immediately suggesting an institutional fix (a comprehensive groundwater management area with various controls) in order to assist in 'rational' decision making, the breadth of solutions was constrained. But the community did not share exactly the same vision of the problem as the government.

Sustainable water management requires innovative processes involving "individual responsibility and informal networks rather than formal hierarchies and control," giving more importance to a shared perception of the problem; planning, natural resource management and decision-making can incorporate "social learning"(Pahl-Wostl 2002, 396). Social learning has many benefits: increased understanding of the resource; identification of common purposes and areas of agreement and disagreement; and an enhanced confidence, capability and sense of citizenry for participants (Muro and Jeffrey 2008, 333).

It appears that the Hornby islanders have followed just such a process. Their collective action captures both factual knowledge and intuitive understanding (as described by Pahl-Wostl 2002):

- Building up a shared perception of the problem through volunteers who collected data, as well as expert studies

- Recognizing individual mental frames and images and how they pertain to decision-making through community meetings

- Recognizing mutual dependencies and interactions with the many involved groups and agencies
- Reflecting on assumptions about the system to be managed, not simply accepting the government's individual proposals
- Reflecting on subjective valuations of what were thought to be the problems or the solutions, and
- Engaging in collective decision and learning processes over a ten year period.

Future Directions and Conclusion

Future groundwater management needs to emphasize sustainability of the resource: gather more and better information, manage groundwater and surface water as one, develop adequate legislative and pricing controls Canada-wide, understand the effect of climate change, and build community capacity.[35] Communities seeking to protect their groundwater can hopefully learn from Hornby Island and combine this information with experience from other areas.[36]

Given groundwater's "invisible" nature, this chapter offers a brief introduction to its formation, extraction, use and governance. The Hornby Island case study focuses on institutions and informal local networks, on the "political" aspects of water management, and on how ideas for groundwater protection develop and change. It shows what can be achieved when people work together: local, academic and government sources contribute and create knowledge; social learning reflects the committed role of participants. Protection measures in the local community plan now encourage sustainability of the resource and balance multiple and competing uses. The Hornby Island Pilot Project illustrates one path to governance measures, but a lengthy one. The provincial government initially sought a comprehensive scheme sensitive to the area for which it was developed, having both voluntary and regulatory actions. More than 16 years later, the local community plan and volunteer efforts, often education-related, give considerable groundwater protection. There is more to come.

Groundwater currently supplies a small portion of Canada's population, but this will change with increasing population. Canada's groundwater management is far from comprehensive. Barriers to adoption include (Karvinen 1994, 21-22; Kohut 1999): lack of jurisdictional coordination between provincial and local governments; lack of public and political attention and awareness; negative public perceptions; public/political fears regarding fees, taxes or restrictions; and lack of mandate and funds for proper documentation. Perhaps governments simply have other priorities and most big-city dwellers are not aware of groundwater. These suggestions seem to ring particularly true in BC where comprehensive groundwater protection is a long time coming due to lack of political attention and fear that new controls would mean new costs to small communities and the agricultural sector.[37] After all, groundwater users have relied on a free (except for well drilling) and hidden resource.

CHAPTER 5: FINDING AN EFFECTIVE WAY TO CURB WATER USE

> I wonder sometimes if there could be a happy medium between my own guilt/fear/paranoia/prudery approach to water and the water-is-plentiful-and-for-wasting philosophy I so commonly observe around me. I'm not sure there is. Maybe the Road to Damascus experience of seeing air, sand and dribble come out of a household tap is essential *(Stenson 2003, 15)*.

What can be done to prevent Stenson's aptly named "water-is-plentiful-and-not-for-wasting" philosophy? There are many calls for change and government investment in research and guidance. Yet comparing twelve countries in the developed world, Canada has the highest levels of municipal water consumption and the lowest pricing (Council of Canadian Academies 2009, 115). A goal of sustainable water use demands an end to exploitation of precious water supplies and avoidance of new infrastructure. So why are many senior governments reluctant to lead with effective pricing and local governments only slowly coming on board with pricing controls?[38] Why are all levels of government happy to promote water conservation and efficiency, counting on responsible individuals to make the difference?

This chapter examines these questions and the many, varied remedies: appropriate charges for water rights, water rentals and treated water supplies, and water efficiency and water conservation measures.[39] It also looks at some more controversial means to control demand or augment water supplies: private-public partnerships, water markets, tradable water rights, and desalination. Chapter 11 discusses water demand management plans.

Water's Worth

Determining water's value is not easy and there is certainly not one "fit for all" formula (Canadian Council Ministers of the Environment 2010, 3). In general, its value is the maximum amount that the user is willing to pay for its use; in practice these estimates are difficult (Briscoe 1996, 5). Mitchell (1984) outlines three approaches: establishing an explicit value to water before it is developed, considering the value added to consumers' products or satisfaction by water use, or basing value on the cost of obtaining and delivering alternative supplies (Mitchell 1984, 33-34). Another study suggests looking at the value of goods and services in actual market situations, valuing water environments through the costs of recreation activities, using willingness-to pay questionnaires, and determining water's value as an input to production (Canadian Council Ministers of the Environment 2010b, 39). For example, an innovative study of Canadian manufacturing firms considers water value finding that "expenditures for water intake, treatment and discharge rarely exceed 1 percent of total costs" (Renzetti 2005, 24). To value water for each of its uses, we need to know about quantity and quality of water sources and cost for the supply and use of water, the latter often dependent on taxpayer-funded projects.

Vancouver's situation illustrates (McNeil 1992, 427). Winter precipitation is high, but summer rainfall can be very low. Storage reservoirs must be built to accommodate periods of natural low flow or drought; for example, on July 30, 2009 a heat wave and evening sprinkler use resulted in the highest ever daily and peak demand. Significant population increases will mean high capital expenditures on storage and distribution, even with conservation measures. Costs levied to supply water may be insufficient to fund future infrastructure. Further, the construction and expansion of reservoirs will affect the quality of the environment; for example, causing the loss of natural spawning habitat and this too comes at a price.

Vancouver's experience gives a practical introduction to the real costs of water supply. In the simplest case, we could regard water as having little or no cost when drawn directly from a stream or an individual well. More often, the provision of water is subject to the following costs:

- capital expenditures (the hardware for the system such as the reservoir or community well, transport to the water treatment plant, treatment to meet quality standards, and the water distribution system to the customers)
- labour (for operating and maintaining the system), and
- resources (the land and raw water needed as inputs into the system—the watersheds and water sources).

These costs should also include the debt incurred to provide for construction of infrastructure such as a dam, the cost should replacement of this structure be needed (as if amortized over a period of time), and a marginal cost (to represent the cost should the existing infrastructure need to be expanded) (Briscoe 1996, 9).

Costs will vary depending on what the end use is and who is managing the resource. In Canada, the authority to charge for water rests with the provinces who may delegate this responsibility to regional or local governments or to specially–created authorities. Whether aboriginal peoples' authority extends to charging for water on their lands appears to be unresolved (Horbulyk 1997, 37). As previously discussed, not all groundwater sources are subject to a licence system and rental charges. But they will certainly be subject to land ownership and well-drilling costs, plus other costs if a community water supply exists. Surface water may be allocated by provincial governments and an administrative fee and rental charge made through a system of licences/permits. Water supplied by a municipality or other purveyor has user charges calculated in various ways.

Domestic water is still very low in cost. Despite new charges introduced by British Columbia in 2009, the annual rent for a surface water licence for domestic use is just $25 per year or $0.60 per 1000 cubic metres, whichever is the greater (BC Environment 2009, 3). Groundwater pricing in Canada ranges from $1.00 to $143.77 per million litres (Christensen and Magwood 2005, 74). In Ontario, users of more than 50,000 litres of water a day pay $750 or $3,000, based on the type of permit (Ontario Environment 2007). Significant industrial and commercial users pay $3.71 for every million litres of water taken. In general, municipalities and other purveyors under-price water, not reflecting the costs of supply.[40] Charges are often based on a flat fee; about one-quarter of the Canadian population pays a flat rate for water use since they do not have meters to measure water use (Renzetti and Busby 2009, 34). As a result, consumption levels are among the highest in the world; Canadians use 329 litres of water per person per day for domestic use (Canada Environment 2011c). The next sections explore options used to change this situation.

Water: A Special Economic Good

There is a preliminary question: is water a public good requiring some form of intervention by government or should water be left to free market forces (Perry et al. 1997, 1)? Water is abundant in some areas, but scarce in others, requiring capture for many uses. Given the number of activities that need water, there are competitive bidders for its use. Water acts as if it is an economic good. The Dublin Statement of the International Conference on Water and the Environment states that: "water has an economic value in all its competing uses and should be recognized as an economic good" (Dublin Conference 1992).[41] Water can be a commodity, but it is also of essential environmental, social, and aesthetic importance; such benefits are difficult to assess on the basis of monetary value and demand ethical awareness. It is perhaps more true to say that water is not a normal economic good but rather, a special economic good (Savenije 2002). Water is a public good that requires intervention at public levels to ensure its sustainability. One intervention takes the form of water demand management.

What is Water Demand Management?

Water quantity is often an illusion. David Schindler, a biologist, expert on watershed ecosystems, and winner of the 1991 Stockholm Water Prize, speaks of this less pleasant truth. Highlighting Alberta's plight, he points to major glaciers on the Rockies' eastern slopes that have lost 25 to 37 percent of their mass in the last century and river flows in summer down 40 to 70 percent of historical values (Schindler 2006). The problems include decline in water quality and in necessary flows for instream uses. Continuing population growth and new water demands will only make this worse. In the past, when considering water demands, water quality and water quantity were looked at separately. Now it is necessary to consider allocation and pollution prevention together, combine quantity and diversity of water supply, and consider recycling, re-use, and public policy measures affecting all these factors (Spulber and Sabbaghi 1994, 3, 13).

To successfully deal with this complexity, many players, from government to the individual, become involved in water demand management. Water demand management is any measure ("technical, economic, administrative, financial or social") that will: "(1) Reduce the quantity or quality of water required to accomplish a specific task; (2) Adjust the nature of the task or the way it is undertaken so that it can be accomplished with less water or with lower quality water;(3) Reduce the loss in quantity or quality of water as it flows from source through use to disposal;(4) Shift the timing of use from peak to off-peak periods; and (5) Increase the water system's ability to serve society during times when water is in short supply" (Brooks 2006, 524). A further definition provides detail: "Demand-side practices include conservation pricing, smart technologies, public education, and regulation that forces innovation by promoting efficiency, conservation and recycling" (Brandes et al. 2005, 5). To understand why these activities are important, we need to determine the value (or cost) of water.

Full Cost Pricing

Pricing of water is an efficient, effective and equitable tool to achieve sustainable use of water, also allaying the cost of water supply. With full-cost pricing,

The consumer will:

- avoid water waste by finding substitutes, changing consumption patterns, and conserving water
- achieve the greatest social benefit from a limited water supply
- reduce water conflicts

Government will:

- build less infrastructure when consumers reduce water use
- generate revenue to cover the cost of water management and new tools
- gain information about value, as opposed to the cost, of water.

The environment will benefit from:

- better water quality
- water conserved that can be re-allocated to instream uses, and
- reduction in environmental costs

    (Pearse 1987 cited in BC Environment, Lands and Parks 1993-2, 7; Renzetti in Bakker (Ed.) 2007, 263); Rogers et al. 2002, 2)

It is undoubtedly true that higher prices may reduce the need for infrastructure, but now there is a new reason to charge appropriately for water. Population increase in urban areas means a need for new facilities and, even while this is happening, the operation and maintenance of ageing water supply and waste treatment systems require new funding. Large urban areas find their taxation powers limited by the Federal government. Senior governments are expected to take a lead role: "Water infrastructure is a unique type of infrastructure: we rely upon it every day, it is critical to our health and to economic development, and it is dependent on, and responsible for supplying, a vulnerable and limited resource that is under increasing pressure" (Canada Infrastructure 2004b, 3).

Funds for infrastructure projects were easily available at one time—appropriate water pricing was not considered. In the 1950s and 1960s, the Ontario Water Resources Commission gave municipalities low interest loans, deferred principal payments for 5 years, and guided infrastructure planning and operation (Shrubsole 2001, 39). In 1969 Ontario began to directly subsidize the building costs for water supply and sewage treatment. Canada Mortgage and Housing Corporation gave capital grants for water infrastructure during the 1970s and, with the Ontario grants, full financing of projects could be achieved.

These programs are again popular. Local governments are eligible for funding of water and wastewater projects under federal initiatives and in partnership with provincial governments. In addition to grants, local areas resort to special district financing. An extra tax is paid by those who benefit; development cost charges make new development pay for services; or less common, a bond issue pays investors an agreed interest over time. Despite these methods, there is only so much funding.

With new challenges, collecting sufficient revenue for water infrastructure is essential. Future water pricing should (BC Environment, Lands and Parks (BC Environment) 1993 - 2, 8; Renzetti in Bakker 2007, 264):

- give sufficient economic incentive to use water prudently and efficiently
- reflect use; the more water used, the higher the price charged
- take into account relative ability to pay
- recoup enough revenue to make water supply operation viable and safe over the long term, and
- reflect the scarcity of the resource, the social cost, the environmental cost, the proportion of water consumed versus returned to the system, and the state of the water returned (whether clean or dirty).

Pricing could also contribute to sustaining aquatic ecosystems and recover the costs of some water uses (such as fire fighting) and water management services (planning, water quality management and flood protection). While the final two suggestions may not be agreeable to all, there is no doubt that for water management organizations having a broad mandate, such costing may be the only means to undertake programs where general taxation revenue is not available.

The above principles are not mere theory. Cochrane, Alberta is taking up the challenge by using "conservation-based" pricing (Town of Cochrane 2006). First, they identified average consumption over 4 years and decided what a fair essential consumption rate would be. They set a goal of decreasing the annual per capita residential water use by 25 percent by January 2009, compared to water use in 2004. This goal would be accomplished through rate increases and a conservation strategy (including education and community outreach). The Town of Cochrane recognized that their rates covered only current operating costs of the water utility; nothing was set aside for future infrastructure upgrades, conservation initiatives, and pollution mitigation. An extensive research study revealed that an appropriate rate structure would reduce municipal water use through an "increasing block rate structure", the most popular method to use. Water consumption has been reduced by 15 percent since 2008 based on the town's water conservation bylaw.

The Town of Cochrane's work is a good start. Ultimately, organizations supplying water need to adopt full-cost pricing, going beyond the cost of supplying water to:

1. Recover all the costs of administration (including actions to plan, inventory, analyze, map, allocate, monitor and regulate water use)

2. Reflect the full value of water use (capture the negative costs to the environment or foregone benefits of alternative uses), and

3. Capture "economic rent" (revenues generated from water use that exceeds the costs of its use including a return on investment or "normal profit") (BC Environment 1993, 10).

Each of these steps could be pursued over time before moving on to the next, but with them, social values and equity considerations need to be taken into account. At its most basic, this means that water should not be disconnected if people cannot pay their bills. Pricing for domestic uses could be set simply to recover the costs of administration, while irrigation (which helps to achieve self-sufficiency in food production) might be subsidized by revenue from other sources (BC Environment 1993, 12). For example, Ontario has introduced a fee for the administration required to issue a "water withdrawals" permit, but has exempted small farm operations from this fee. The structure used for pricing water appears to be more important than the price level in effecting change in water demand (Reynaud and Renzetti cited in Cantin et al. 2005, 4). A more appropriate pricing structure can result based on varying the tariff elements.[42]

Other factors can be taken into account such as charging more where water is scarce or charging the same price for similar uses no matter where the location. We can adjust water rates over time to reflect seasonal scarcity and charge more for water with altered quality or not returned to a water source (BC Environment 1993, 8). Such innovative charging systems have not been the norm. Most water users are subject to a flat rate up to a designated quantity or a fixed charge, after which a constant rate per unit takes effect. This section presents facts about water pricing as an important means to control water demand. The next section demonstrates problems with this approach, before moving to other water demand management measures frequently adopted as panacean.

Why Not Increase the Price of Water?

Of the reasons for rejecting pricing as a water demand management tool, surely the most common is the political storm caused by any increase in fees. Strong negative reaction follows when the major effect is on those least able to pay, or if, for those who can pay, no change in behaviour results. Additionally, there is an apparent public blindness to the costs of any project supplying a public good such as water, where the costs are borne nationally or provincially. Why wouldn't people, particularly the development community, be prepared to ignore the costs if a brand new facility for the provision of more drinking water is the result? And then there is the dilemma of how to promote change within organizations used to thinking in terms of water supply provision rather than water demand management. Fortunately, the latter situation appears to be fading rapidly.

There is also an impediment to volume pricing in the need to measure use by metering; many land uses, including high density residential areas, are not metered (about one-third of households are not metered) (Renzetti and Busby 2009, 32). The cost of installing meters is cited. Even though benefits may outweigh the costs and metering is fundamental to the introduction of pricing controls, progress is slow. Starland, Alberta reacted early to water shortages by implementing a public education program. The district mailed more than 700 pre-packaged water conservation kits to households. In the kits were dye-testing tablets to detect leaks in toilet flappers, water saving devices for bathroom and kitchen faucets, a low flow showerhead, a tap aerator and an information card. Over 90 percent of the residents (population of 2,210 in 2001) reported installing the kits. Later, 84 residents installed meters (Canadian Water and Wastewater Association 2004).

Finally, there is an unintended result to increasing water pricing to promote conservation. BC's Capital Regional District has slowly increased water prices over the last ten years. In 2010, water prices rose by 15 to 20 percent to cover the shortfall in revenue caused by declining consumption (Clarke 2010, S3). For the same reason, the City of Winnipeg raised water prices by 3.2 percent in 2010 (Canadian Broadcasting Corporation 2010b). Given the wariness which governments display to the displeasure of their taxpayers, water efficiency measures are often more attractive than making changes to water pricing. This alternative was supported by 24 water demand management experts, despite acknowledgement that: "the lack of an appropriate pricing structure significantly limit[s] the potential for successful water demand management in Canada" (Maas 2003, 5, 17). Seven years after the publication of their above-noted report, the POLIS organization at the University of Victoria is now promoting water pricing (Brandes et al. 2010).

Water Efficiency and Conservation Measures

Water efficiency and water conservation measures, promoted by government and implemented by communities and individuals, are often seen as more acceptable than changes to water pricing. The emphasis is on behavioural change; reducing the amount of water we use and the rate at which we use it in a number of different ways. This section briefly explores these tools.

In Canada, many agencies have studied and promoted water efficiency measures. The Science Council of Canada was an early leader. They encouraged recycling of cooling water, reusing water in industrial processes, detecting and repairing leaks and infiltration in water systems, lining irrigation canals to reduce seepage, delivering water to crop need, and using xeriscaping (Brooks and Peters 1988, 12ff).[43] Environment Canada's *Water: No Time To Waste—A Consumer's Guide to Water Conservation* (1995) brought hints for the home with sections titled "In the Kitchen", "In the Bathroom", "In the Utility Room", "In the Outdoors", and a section at the end called "Water Log". Seeing how your household measures up brings a resounding surprise.

In 2000, the Canada Mortgage and Housing Corporation (CMHC), in partnership with local areas, produced a *Household Guide to Water Efficiency*. The Utility Room became the Laundry Room in this brochure and information, coupled with beautiful illustrations, explains water efficient landscaping and installation of low flow devices (CMHC 2000). Since then, education efforts have increased exponentially. There are many useful sources (Brandes et al. 2005; Roach et al. 2004; Shrubsole 2001) and government agencies, institutions, the private sector, individuals and the media have much to contribute.

The Federal government promotes water efficiency and conservation through general guidelines and the study of water's value (Canadian Council of Ministers of the Environment 1994, 2010b), and by promoting the development of new technology. Industry Canada's Canadian Environmental Solutions supplies a directory of Canadian companies where new developments can be found in everything from plumbing fixtures to automatic meters. Agriculture and Agri-Food Canada fund projects using best management practices.

Provincial governments encourage conservation through promotion (Council of the Federation), legislation (Ontario's 2010 *Water Opportunities and Water Conservation Act* allows the Minister to set targets in relation to the conservation of water), allocation decisions, regulation of water use by local governments, building and plumbing codes containing the most appropriate standards, and best practices in irrigation of agricultural areas. Infrastructure funding from both levels of government can be made dependent on water efficiency plans. In drought[44], more serious measures are necessary. BC's Okanagan region was subject to a severe drought in 2003. The provincial government acted quickly to ensure better planning in the future (BC Land and Water BC 2004). A Drought Index and Drought Management Plans encourage management teams for local areas to know their water supplies, improve water use efficiency, and undertake communication and education efforts.

Local (and regional) governments may stipulate sprinkling restrictions, requirements for water-saving fixtures in new construction and renovations, use of grey water (recycling of wastewater), and collection of rain in rain barrels. They develop water conservation plans and give rebates for the purchase of water efficient equipment. Dauncey (2004, 6) suggests that local areas pass a bylaw that limits building permits for new houses until enough existing houses are refitted to save the necessary water for the new construction. All governments can ensure that their own buildings and recreation areas are using the latest in water-conserving equipment, watering parks and landscaping in a responsible manner, using xeriscaping and shrubs that capture rainwater runoff where possible.

Institutions also have a role to play. At the University of Victoria, as part of their sustainability initiative, treated water leaving the Outdoor Aquatic Facility is recycled into new buildings to be used in toilets and urinals. In addition, toilets in older buildings are upgraded to low-flush fixtures that use less than 60 percent of the water consumed by traditional toilets (University of Victoria Facilities Management 2004, 6).

Water suppliers need to monitor for leaks in their distribution system, including leaks in systems of their customers, and promote water conservation. And fix those leaks! The City of Montréal waterworks were reported to leak more water than they deliver to residents (Yakabuski 2005, A3) and in January 2013 an 88-year old water main burst, causing serious flooding in the downtown area. Costs to fix Montréal's crumbling drinking-water and sewer systems are estimated at $4 billion over 20 years (Minardi 2010, 8).

Agriculture accounts for 10 percent of gross water use and 66 percent of water consumption (mainly crop irrigation) in Canada (National Round Table 2010, 19). Besides the quantity implications of over-use, groundwater quality may be affected and ecological damage occur (Canada Policy Research Initiative 2005, 8). While pricing reform does not appear to affect demand to a significant degree, irrigation districts and improvement districts can play an important role when given assistance to acquire new irrigation technologies and upgrade systems (Canada PRI 2005, 31).

Industry participates through water efficient initiatives such as using closed-loop systems and air cooling in thermal power and cooling both by retrofitting and in new operations, reducing water withdrawals in all industries, and substituting saline water or recycled water for the use of fresh water or steam in oil extraction and bitumen operations (Kinkhead 2005, 3-4). The municipal development sector can encourage green building projects; for example, the City of Victoria's innovative Dockside Green project treats water and collects rainwater to be used to flush toilets and to supply irrigation for ponds and watering. Hopefully this trend will continue based on Health Canada's *Canadian Guidelines for Domestic Reclaimed Water for Use in Toilet and Urinal Flushing* (Working Group 2010).

Education about water conservation is one of the most powerful tools available, at the least cost. The City of Durham, Ontario (Brandes 2005, 20) and the Capital Regional District in Victoria have hired students who work with home owners to encourage conservation practices. The media play a significant role by alerting the public to water use statistics and regulations regarding water use. The online multimedia portal, water.ca, extends the reach of education efforts. As for all aspects of water management, success depends on the notion that once you understand why it is important to behave in a certain way, then you will do the right thing.

Which is Better?

There is no clear answer yet as to the better path, using pricing or adopting other water efficiency and conservation measures. Future research will give a more appropriate comparison. But a *National Action Plan* to encourage municipal water use efficiency acknowledges that Canada is behind other countries in providing regulations and policies including available conservation measures and economic instruments (Canadian Council Ministers Environment 1994; Canada Infrastructure 2004a, 15). While we need leadership from federal and provincial government agencies (Brandes et al. 2005, ii), success stories suggest that much will depend on local government, water purveyors, and individuals.

Kelowna, BC installed meters in 1997, implemented a constant rate structure based on metered consumption in 1998, and full cost-recovery rates in 2000. By 2001, average residential water use was reduced by 19 percent. Due to a semi-arid climate and poor soil conditions, Kelowna needed to take more action in the summer (Webb 2004, 34). The city's Water Smart staff identified neighbourhoods with the highest water consumption and offered incentives to reduce outdoor water use: lawn aeration and top dressing (composed bio-solids from the wastewater treatment program), professional irrigation system assessment, cost-shared improvements and customer education. For July 2001 water use decreased an average of 26 percent from the previous year (with weather factored out of the results). Education was the most effective initiative. Subsequent programs in other neighbourhoods show the importance of tailoring the program to the specific area (Webb 2004, 26).

Actions to Encourage Water Demand Management and Augment Supplies

Other government/private sector actions encourage water demand management or augment water supplies; their possible costs and benefits should be considered. First, the private sector may be involved in building and/or operating infrastructure and in so doing become a leader in water efficiency measures or in full-cost water pricing. For example, the corporation EPCOR provides water services to the City of Edmonton. The City transferred ownership, operations and maintenance of its water supply assets to the corporation, but retained responsibility for wastewater and drainage. EPCOR needed to reduce peak water consumption or build a water treatment plant expansion, inevitably leading to water price increases. As water was already highly priced, the company implemented a 15 year conservation program: full metering and a 'Train the Trainer' program to educate plumbing apprentices, hardware and plumbing suppliers, building owners and developers about the benefits of ultra-low-flush (6-litre) toilets. Public education focused on a rain catchment program, Odd/Even lawn watering, and use of non-potable water. Of the information and education programs, the Train the Trainer Program has been most effective. The initial goal to reduce water use by 10 percent in 8 years was exceeded (Canada Mortgage and Housing Corporation, accessed 2007).

Use of the second action, water markets, is infrequent in Canada and is more applicable to the management of water in local areas. Water markets are useful where a system of prior allocation exists. Riparian rights are tied to the land and thus are not marketable. Water markets encourage the use of water rights' trading from low to high value uses. This has the potential to create a more efficient allocation of scarce water resources, while at the same time allowing governments to reclaim water flows for protection of the environment.

Those who believe in the power of markets are passionate in their views: "Growing demands ... will put pressure on limited water resources. Those pressures need not create water crises if individuals are allowed to respond through market processes. Some would say that water cannot be entrusted to markets because it is a necessity of life. To the contrary, because it is a necessity of life, it is so precious that it must be entrusted to the discipline of markets that ensure conservation and innovation" (Anderson and Snyder 1997, 204). Market systems, most notably used in Chile and Australia, depend strongly on a well-developed policy framework including a water rights system and a management authority able to protect ecological needs. The system must take into account return flows (not used by licensees) to rivers and changing land use (Economist 2003b).

Alberta enables water transfers.[45] Transfers may be on either a permanent or temporary basis, require an approved management plan, and benefit from the government's ability to withhold 10 percent of the water for environmental purposes (Yee in Cantin et al. 2005, 6). Transfers can involve the full allocation of a water rights holder or a fraction of it. Transfers can be sold or leased for a fixed period of time and buyers and sellers must find each other. While maintaining the priority of the existing allocation, the transfer relocates water to the new location within the basin (Palacios 2005, 25). A model of the Alberta system suggests that such a cautious approach will result in relatively large efficiency improvements in drought years (Mahan et al. 2002, 47).

The Alberta system came under review after a developer agreed to pay an Irrigation District $15 million (for water-saving measures) to transfer water from the Balzac River to service a shopping mall and racetrack under construction near Calgary (Canadian Broadcasting Corporation 2007a, Christensen and Droitsch 2008, 18). Subsequent recommendations from the Alberta Water Council (2009) support the concept as legislated. However, the Council calls for water to be set aside for environmental and non-consumptive purposes before implementing the transfer system and a "robust" and transparent market with clear guidelines. In addition, the Council recommends that gains from conservation or efficiency be made available to the market. Operational improvements suggested include plans for water conservation and water shortage, an open application process for more scrutiny in certain circumstances, and an accessible information platform.

Resistance to the concept of tradable water rights comes from those who believe that water should not be sold, that water is a human right. Concern focuses on the belief that rich companies will buy up all the water and then raise water prices, excluding others from a basic right. New rights holders might over-use their allocation or return polluted water to water bodies. There is the possibility of damage to third parties where water is used in an urban area to the detriment of the future of small communities, agriculture, recreational areas or the common good. In Canada, as in other countries, there are also implications of water transfer schemes to future aboriginal land claims.

Finally, another approach, communities in coastal areas might turn to desalination to avoid using fresh water at all. Infrequently used in Canada except by small users (for example, by owners of small islands in the Gulf Islands area of British Columbia), desalination would only become really practical when the cost of the process is lower than the costs of other water supply alternatives. Desalination occurs by distillation; heated salty water produces vapour and then condenses to produce fresh water. Reverse osmosis involves forcing water through a membrane at great pressure leaving the salt behind. A Canadian company, Zenon, started in Hamilton, Ontario, has become a leader in this field.

Future Directions and Conclusion

Successful water demand management depends on political leadership, evaluation of programs, effective control of private sector initiatives, new planning initiatives (see Chapter 11), and public involvement and self-monitoring. The University of British Columbia's Program on Water Governance calls for a national strategy and greater provincial direction to ensure municipal accountability (Furlong and Bakker 2008, 37-38). The Program further suggests efficiency requirements for water-using fixtures, provincially-enabled programs for municipal water conservation and regulation, full-cost recovery legislation and approval of financial plans, and tying water allocation to water-use efficiency (Furlong and Bakker 2008).

Setting goals, benchmarks and targets and formally measuring performance (indicator data and recording of ambient data such as price, precipitation and temperature) are necessary (Webb 2004, 25-28). While there is a wealth of information on water efficiency and conservation measures, less is available on asset management (particularly water pricing and metering), the state of infrastructure, and the effect of public involvement on infrastructure decisions (Canada Infrastructure 2006). The use of market-based instruments needs more research (Horbulyk 2005);

The role of the private sector will continue to be challenged. While there is criticism of public-private partnerships for water supply and other systems, this trend is likely to grow. To ensure continuation of positive aspects of these initiatives and retain public trust, government needs to emphasize results-based management and transparency.

Greater community participation in decisions about water supply enhances understanding about the need to control water demand. Large organizations are more likely to have the necessary resources and techniques to promote this than small utilities, so suitable tools need to be developed for the latter. Education programs based on protection of the ecosystem are essential. As the Canada West Foundation suggests: "urban water users ... may be disconnected from the broader environment and unaware of their impact on it" (Roach 2004, 12).

Comprehensive planning for water demand management shows promise in some areas (see Chapter 11), but for smaller communities, it is better to tailor water efficiency initiatives to particular situations (Webb 2004, 36). For the individual, there are responsibilities beyond using water more efficiently. At the household or commercial enterprise level, the concept of self-monitoring will grow in importance in areas where no metering exists. Relating to the larger picture, we need to scrutinize borrowing for infrastructure, the use of public private partnerships, and water transfers.

This chapter shows why controlling water demand takes time. Full-cost pricing appears to be the most effective way to discourage water use and fund necessary projects, but there is a perceived political disincentive. Governments have continuously adopted persuasive actions, sometimes followed, at the local level, by "soft" regulation such as watering restrictions. Institutions, businesses and individuals have begun to take responsibility for water-saving.

In earlier times government-funded projects supplied all the water required. Now many factors (population increase, new knowledge about drinking water quality, ageing infrastructure, and climate change) lead to new attempts to control demand. Water demand management efforts cannot be just the "flavour of the month". Governments need to show leadership: commit the resources necessary; encourage interdisciplinary studies with emphasis on valuing water both for use and non-use; introduce realistic water pricing to support future infrastructure, conservation programs, and ecosystem protection; adopt appropriate oversight where privatization occurs; and maintain consistent vigilance to ensure future progress in their efforts. Individual efforts, whether guided by government, or not, are paramount.

THEME 2: PROTECTING PUBLIC SAFETY AND WATER'S ECOLOGICAL AND INTRINSIC VALUES

CHAPTER 6: WHO SPEAKS FOR THE FISH AND WHO FOR THE STREAM?

This respect for the fish's environment ... implies not merely a concern for such things and a desire to protect them, but a positive affection for the whole natural world and a deep desire to understand it *(Haig-Brown 1964, 182).*

At the basic level of sensory experience, water appeals to the whole: it can be seen, felt, smelled, touched, and tasted *(Buttimer 1984, 263).*

The title to this chapter poses questions that are often asked, but answers can be elusive. This chapter seeks response through protection of instream flows[46] (including those used when water is consumed without prior treatment), riparian areas, and the intangible values of water. The next chapter discusses protection for safe, secure drinking water supplies. Both are subject to many of the same problems. Allocating too much water (introduced in Chapter 3), disturbing riparian areas, or allowing pollution to occur affects public health and safety, the environment, and the economy. And the prognosis is not good. In 2009, a meeting of the Instream Flow Council, representing state and provincial fish and wildlife management agencies' instream flow programs, comments: "the ecological function of streams and lakes of Canada and the U.S. will probably stay the same or decline" (Annear et al. 2009, 72). Rob de Loë, University of Waterloo's Research Chair of Water Policy and Governance suggests that ecosystem protection is "still in its infancy" (de Loë et al 2007, 45).

Problems Affecting the Aquatic Ecosystem

Instream water and streamside vegetation are crucial to the aquatic ecosystem. Fish and other aquatic organisms require living space, cover in the form of streamside vegetation and organic debris in the stream, a continuous supply of oxygen, and organisms and vegetation for food supplies (that also require water to survive) (Hatfield and Smith 1985). Fish require special flows, unblocked access to spawning sites, and a stream bed having clean unsilted gravel beds for spawning, incubation and the growth of new fish. In addition, there must be adequate flows to allow movement and migration of fish, as well as specific water temperatures. Good water quality and appropriate turbidity[47] are also important. Besides the needs of fish, beaver, muskrats, waterfowl, frogs and toads require water and riparian vegetation. The vegetation gives cover and food material and maintains water quality by securing sediments. In addition, instream flows may be needed to maintain groundwater levels and to sufficiently dilute pollution.

Problems affecting the aquatic ecosystem are now an almost daily media focus; current topics include the effect of the Alberta tar sands development and loss of wilderness areas. Reading these reports, it's too easy to ignore the context in which they occur, the watershed, just as it is difficult to come to terms with the complexity that results when problems become apparent. With this in mind, this section adopts a watershed as the base for its discussion.[48] Surface water and groundwater, as they occur in nature, may contain elements that make them unsafe. Obvious examples include the presence of silt, minerals, and vegetation, and salt water intrusion in coastal areas. But there are a plethora of other problems.

Over a watershed area, environmental developments affect water sources. Acid rain causes a decline in the pH of water bodies, extinction of fish, and reduction in forest growth protecting water bodies.[49] Climate change can lead to an increase or decrease in the amount and variability of rainfall, so affecting water quantity. Soils that erode through natural or human-induced events (landslides, forest fires or forestry) cause turbidity. These are non-point sources of pollution, one or several activities occurring over a broad area, as distinguished from point sources: single, identifiable sources (e.g., waste emanating from a factory or treatment plant). Several examples illustrate.

Logging, especially clear cuts, and improperly constructed and decommissioned logging roads disturb water quality, aesthetic values, and habitat. Dirt-biking in watershed recreation areas can be "a recipe for disaster" when the resulting erosion plugs culverts and disturbs soil, washing it into the creeks that supply water (Hume 2005, S3). But nature is the bigger culprit when heavy rain causes mudslides that wash into rivers and lakes.

Despite efforts to prevent the deleterious effects of mining and oil extraction or to clean up after these operations, threats remain from over-extraction of water and contamination in the form of atmospheric emissions, drainage, tailings disposal, spills and discharge water. Serious acid mine drainage problems exist, particularly in mines active prior to environmental regulation, perhaps as many as 10,000 mines across Canada (Mining Watch 2005). Open pit coal mines now raise concern. International opposition arose over the Cline Mining Corporation's proposal to develop a mine 22 miles north of Glacier National Park in BC's Flathead Valley, a valley renowned for its wildlife diversity, number of vascular plant species and the pureness of its water (Quinn 2007, 40). Montana residents in opposition cited possible release of toxic heavy metals or other pollutants into the North Fork River and fear that even more mining operations would follow (Peterson 2006, 1). A Memorandum of Understanding between BC and the US was signed in 2010 that denies mining, oil and gas development and coal-bed gas extraction in the Flathead Valley and the Nature Conservancies of Canada and the US have secured $10 million in funding to assist implementation of this agreement.

Over-extraction of water and pollution caused by oil and coal-bed methane drilling for natural gas is a significant current issue; myriad well-researched reports and media attention have caused embarrassment to Alberta's oil sands industry and the government. "During the past year a variety of industry and government agencies have recognized that the intensive water requirements of unconventional oil, combined with climate change, may threaten the water security of two northern territories, 300,000 aboriginal people and Canada's largest watershed: the Mackenzie River Basin" (University of Alberta and University of Toronto's Munk Centre 2007, i).

Sixty-six percent of the licensed surface water allocations from the Athabasca River and its tributaries are committed (Gordon Foundation 2006, 5; Nikiforuk 2008, 46-52). Dr. David Schindler presents more chilling facts. Of the water used, 8 percent is returned to the river; the remainder is too polluted with carcinogens, arsenic, mercury, lead and other metals. This water goes to tailing ponds that could be described as "tailings lakes". These ponds show leakage and have the potential to break. Should breakage occur in winter, it would be impossible to mop-up (Schindler 2010).

These concerns led Canada's Environment Minister to create an expert panel to study impacts of toxic elements released into the Athabasca River (Kelly et al. 2010). The Royal Society of Canada's report *Environmental and Health Impacts of Canada's Oil Sands Industry* (December 2010) calls into question the regulatory capacity and environmental impact processes of the federal and Alberta governments. In response, these governments vow to establish "gold-standard" monitoring (McCarthy 2010b, A13). In February 2012, the *Alberta Environmental Monitoring Panel* began work to establish an agency to lead monitoring, evaluation and reporting and to find a stable source of funding for these activities.

Also in 2012/13, public protests in New Brunswick highlighted the issue of shale gas fracking, a process by which water and other substances under pressure are injected into shale rock formations or coal beds to release natural gas when fissures in rock are widened. The problem is that this process uses large amounts of water, requires significant space that may disturb habitats to store waste water, and includes chemicals that can contaminate surface and ground water (Canadian Broadcasting Corporation 2012). In response, industry is setting operating principles, at least a start, but these only begin to address both water quantity and quality concerns as use of fracking becomes more common.

In a watershed, chemical spills can occur, for example, from a derailed railway tank car. A spill of 51,000 litres of caustic soda into the Cheakamus River near Squamish, BC killed up to 95 percent of the river's fish and resulted in a short-term ban on drinking water from the river and wells within 100 feet of the river (Mickleburgh 2005b, S1, S2).[50] More recently, the BC government has seized a site polluted by a gas station's fuel tanks. Leaks seeped into one of Vancouver Island's most important fishing sources, Oliver Creek; clean-up will cost $1 million (Hume 2008a, S1).

Agricultural contamination of water results from animal grazing and feedlots, manure piles, fuel and fertilizer storage and land application of fertilizers and pesticides. In July 2007, over a thousand rainbow trout and salmon were found dead and thousands of others were affected, washed up on the banks of Prince Edward Island's Tryon and Dunk Rivers, probably the result of pesticide runoff (Tutton 2007, A7). Erosion, leaching, and draining of wet soils carry pollutants (sediment, nutrients, pesticides, bacteria and salts) into receiving waters. For example, some crops need nitrogen fertilizer. The high residual nitrogen levels that result are apparent in "the lower Fraser Valley of British Columbia; agricultural land from Lethbridge through Red Deer to Edmonton in Alberta; the Melfort area in northeastern Saskatchewan; the Red River Valley in Manitoba; southwestern Ontario, the area around Lake Simcoe, and the lower Ottawa Valley; the St. Lawrence Lowlands in Québec and the region south of Québec City; the Annapolis Valley in Nova Scotia; and the St. John River Valley in New Brunswick (Coote and Gregorich 2000, 1, 37)".

Other problems are evident as land use in the watershed intensifies further downstream. Industrial sources such as pulp mills may produce toxic air pollution or, as occurred in the past, dump neurotoxins into watercourses. Digging in gravel pits can cause sediment to wash into streams and contaminants to leach into groundwater and surface water from acid-generating waste rock. Urban and industrial effluents present the most serious problem; in the area surrounding the Great Lakes "many cities … have antiquated systems for collecting and treating sewage and regularly release untreated sewage into local waterways" (Sierra Legal (now Ecojustice) 2006). Their actions rely on "assimilative capacity", the ability of natural water sources to absorb, dilute, and disperse wastes. Today, we are aware that the discharge of waste requires control, but even these efforts generate problems. Sewage sludge, containing valuable nitrogen or phosphorus, may also include heavy metals and toxic organic compounds.

Leachate produced in garbage landfill sites will damage water quality. By 2015, end-of-life batteries requiring disposal will exceed 20,000 tons (Canada Environment 2010a, 5). These batteries may contain heavy metals that can leach into water sources. More difficult to trace are sewage overflows from sanitary sewage systems that mix with storm water and enter the underground sewer systems during significant storm events.

Urban development results in sediment, toxins and pathogens being washed into water bodies. Paved surfaces prevent infiltration of rainwater into groundwater and accelerate runoff into rivers, streams and lakes. The amount of stormwater can increase dramatically in flashy storm events, now common and attributed to climate change. Contaminants in stormwater may include toxic chemicals (gasoline, pesticides, oil, grease, metals and detergents as well as other solid items such as cigarette butts and trash) and fecal coliforms (Capital Regional District Environmental Services 2007). Even leisure pursuits bring problems; treatment of golf courses with fertilizers and pesticides is an example. For Calgary's Bow River, the runoff from ten golf courses illustrates the issue (Cullen 2003, A12) and Armstrong et al.'s (2009) *The River Returns* tells the environmental history of the river including earlier problems. One of these was "the Blob" that appeared in 1989, residue from an abandoned wood-preserving plant in the river bed, not found during an earlier clean-up.

There are other pollution and contamination sources. Hydraulic works for flow regulation, electrical generation, irrigation and flood control can have unintended consequences. With damming of the Nelson River, and recent flooding events carrying greater loads of phosphorus and nitrogen, Lake Winnipeg has become a "giant holding tank, retaining the nutrients much longer and allowing them to feed algae blooms"; the algae are gradually suffocating the lake (Ferguson 2004, F6). Algae are fed by nitrogen and the nutrients in manure, sewage and industrial waste, mainly from agricultural operations such as hog production (Schindler et al. 2008).[51] Chapter 9 discusses other water quality problems relating to dam construction.

Zebra mussels and other invasive species released from the ballast water of freighters now affect the Great Lakes and other smaller lakes (see Brooymans 2011, 211). Zebra mussels clog water intake pipes and their growth may provide habitat for avian botulism, toxic algae and toxic chemical contamination (Mittelstaedt 2002, F1). Invasive species do an estimated $200 million (U.S.) damage per year

(Anderssen 2009, F6).[52] Pleasure boating may not be the most obvious source of water pollution, but people often release untreated sewage carrying parasites, bacteria and viruses into fresh water; toxins from paints and wood preservatives cause tumours and lesions in certain fish and other aquatic life. As well, boats may leak fuel and oil. The problems described in this section may affect the aquatic ecosystem and our ability to enjoy water for its own sake. They may also affect drinking water quality. This has led to extensive protection efforts.

Aquatic Ecosystem Protection

There is an inextricable link between the ecosystem and water: "A river's flow regime is the single most important determinant of the river's ecologic condition" (MacDonnell 2009, 1087). Groundwater is also important as its springs feed creeks and create fish habitat. Any discussion of water protection starts from this point. While maintaining more water instream does not necessarily guarantee more fish, most resource managers have come to take this position given their experience of over-allocation for other uses and mistakes in determining the flows necessary for fish protection (Hatfield and Smith 1985, 1).

Damage to instream flows is the result of small local activities and significant major ones: over-allocation of water licences, historic waste disposal practices, and the construction of major projects such as the Churchill River Diversion in northern Manitoba, the James Bay project in Québec, the Churchill Falls hydro development in Labrador and the Nechako-Kemano Diversion in northern British Columbia. Measures to protect the aquatic ecosystem have developed over time and now include a patchwork of government and non-government activities. The recent past has seen considerable activity, particularly at the local and provincial government levels, making up for earlier resistance or lack of awareness.

Federal Government

The Federal government's involvement in instream flow protection is limited. The *Canadian Environmental Protection Act* (1999) provides control against toxic substances entering the environment; for example, the concentration of phosphorus in cleaning products (a 2010 amendment). *The Fisheries Act* makes it an offence to deposit deleterious substances into water frequented by anadromous fish[53] or to alter, disrupt or destroy fish habitat. This authority is affected by a decision of the Supreme Court; federal powers do not extend to the general regulation of water quality, but simply to any real or apprehended harm to the fish habitat (*Fowler v. The Queen*, [1980] 2.S.C.R. 213.) (Irvine 2002, 12). These provisions are further weakened by 2012 amendments (Div.5, Part 3) that narrow the Act to protection of fish habitat that supports commercial, recreational or Aboriginal fisheries.

Fisheries Act provisions are augmented by 2012 amendments relating to works in and about streams and deposit of deleterious substances. The *Great Lakes Basin Compact* commits individual signees to maintain water levels through regulation of water withdrawals. Although authority exists through the *Boundary Waters Treaty* for protection against point-source pollution, provincial and territorial governments issue permits (Muldoon and McClenaghan 2007, 249). Where the Federal Department of Fisheries and Oceans has been very effective is in their publication of guidance to aquatic ecosystem protection (Canada Fisheries and Oceans 1993/94).

Provincial Government

Other than the general ability to reserve water, provincial protection of instream flows was virtually non-existent in 1986 (Reiser et al. 1989). By 2009, most provinces had a legislative commitment to conserve and protect the environment (MacDonnell 2009, 1084), but protection often relies on the presence of a licensing or permitting scheme as the opportunity to consider fish and other ecological values. Unfortunately, it is not always possible to issue a licence or permit to volunteer groups who do not have sufficient funding to maintain the water pipeline and the Department of Fisheries and Oceans is no longer willing to assume this responsibility given liability concerns (Hume 2011a, S2). Some provinces specifically acknowledge an environmental purpose as a beneficial use in a licence. But the uses of instream flow are usually expressed in an anthropogenic context, associated with some form of water right.[54] To protect instream flows, the provinces have various provisions[55] Alberta allows transfer of water rights as previously discussed. An innovative example is the application by ConocoPhillips to donate 50 percent of its licence on the Medicine River to the Water Conservation Trust of Canada to protect instream flows (Water Canada 2010).

Provincial involvement in instream flow protection is hampered by the need to first identify the stream flows and water quality required, a costly and time-consuming process. These calculations are complex including temperature, dissolved oxygen, and turbidity; hydrology and the amount of water present at a given time; the biological systems; and geomorphology. Then legislative and policy mechanisms can come into play, followed up by monitoring and enforcement.

Table 6-1 summarizes some ways to protect water quantity and quality at the provincial level. Pollution control laws and mitigation are essential; for example, committing $4.5 million to stop pollution from an abandoned copper mine in the hope of returning salmon, trout and steelhead stocks (Hume 2008b, S2). There is also the possibility of water and riparian area protection regulations, designation of "sensitive streams" to permit setbacks for development, and the creation of recovery plans. Forestry and range practices and the use of fertilizers, pesticides, and herbicides may be controlled. In BC, "results-based" forestry legislation governs the preparation of plans to protect community water supplies on Crown lands and watersheds with significant downstream fisheries values (University of BC Forestry 2002, 12).[56] At the strategic level, the Northwest Territories water strategy strives for "sustain[ed] river flows, aquatic ecosystem health, and the ecosystem services that make life on earth habitable for all living organisms while simultaneously protecting Northern cultures and important aspects of traditional ways of life in the midst of a changing climate and rapidly expanding petroleum and mining industries in the North" (Baltutis 2012, 6).

Table 6-1: Possible Tools for Provincial-Level Aquatic Ecosystem Protection

Licence or permit provisions:
Reserve water for instream flow, in government's name, with priority over other uses.
Grant licence for instream flow with a priority date; instream use for fish becomes a "beneficial use".
Where water rights transferred, tenure renewed or application changed, reserve instream flows.
Where water rights transfer from an off-stream use, reserve amount for instream use and grant new licence.

Protection regulations other than pollution control:
Regulations specify the time of year for acceptable changes in and about water bodies and wetlands, minimum flow requirements when work may be undertaken, salvage or protection of fish or wildlife during changes, restoration requirements, and protection for water quality (limit sediment deposit and disturbance of natural materials and vegetation).

Water Allocation Plans:
Set aside a specific quantity of flow to be protected against other appropriation.

Other:
Prohibit new dams on protected rivers. Modify reservoirs to improve streamflow. Designate "sensitive streams" to permit setbacks for development, the creation of recovery plans, and riparian area regulations.

Forestry legislation (conditions applied by prescription or through plan provisions):
Watershed assessment evaluates the hydrological conditions and nature of the watershed to determine sensitivity to future development.

Forestry practices: leave buffer strips alongside surface water sources and wetlands; proper construction, design, and decommissioning of roads, skid trails and landings; and post-forestry soil erosion control. Avoid clear-cutting and use sustainable cutting plans. Watershed plans include terrain stability mapping and plan road construction and cut block layouts. Monitoring and contingency plans are developed.

Local Government

Community land use planning has an impressive array of measures to protect the aquatic ecosystem, particularly riparian areas (Table 6- 2). Official Community Plans may contain environmental policies. These over-riding principles and objectives translate into land use designations and specific objectives to protect water bodies. Objectives could discourage certain land uses, major roads and utility networks; specify large lot sizes in environmentally sensitive areas and setbacks from watercourses; and protect wetlands and other groundwater recharge areas. Soil erosion, sedimentation, and work in and about streams can be controlled to maintain fish passage and prevent deleterious substances from entering water.

Some official community plans introduce special tools such as development permit areas, density bonus zones or comprehensive development areas. Where development sites are large and contain a number of uses, comprehensive development areas protect the ecosystem through provisions negotiated prior to development. Screening or landscaping to mask or separate uses from areas of the natural environment can be required. Development permits specify how to protect, restore and enhance riparian areas and control certain activities: removal of vegetation and trees, installation of septic tanks, and even subdivision to create additional lots. These permits may also activate environmental assessment of the proposed development, the provision of leave areas, control of erosion and sedimentation, stormwater management, control of instream work, vegetation management, and bonding of construction firms and environmental monitors (Canada Fisheries and Oceans 1993/94, 25).

Table 6-2: Local-Level Land Use Planning Tools to Protect the Aquatic Ecosystem

Plan provisions:
Map environmentally sensitive areas/designate land for uses that preserve hydrological features.
Include protection goals and policies.
Retain land for parks and recreation sites including walkways around lakes and beside rivers. Enable development permit areas, density bonus zones and comprehensive development zones to trigger protection of the aquatic ecosystem and groundwater recharge. Require land developers to provide environmental information at time of development/enable environmental impact assessment.

Zoning conditions:
Require large lot sizes in critical water protection areas. Require tree-cutting permits; soil removal and deposition control; watercourse protection (no fouling or impeding); and setbacks for floodplain management purposes.
Regulate disposal of storm water or surface runoff from private property.

Subdivision approval:
Dedicate areas for preservation, ideally as part of a greenway system.
As a voluntary act by land owner, request conservation easements adjacent to water bodies including: bequests, donations, sale or lease of lands to a conservation agency; conservation covenants (voluntary written agreements between the landowner and a conservation organization or government agency in which the owner promises to protect all or part of their property); management agreements (voluntary agreement providing for assistance from government in the form of advice or management); profit à prendre agreements (owner assigns rights to trees on a property to a government agency or another individual to ensure that clear-cutting cannot occur) (Islands Trust Fund c1998).

A density bonus is a concept normally used to achieve community improvements and public benefits; it may be used to protect sensitive ecosystem areas. Density bonusing allows zoning requirements to vary in exchange for beneficial amenities specified in a community plan. Developers may build more floor area on a reduced land area, from which they derive more income. In return they provide an amenity such as environmentally sensitive land at the edge of a river. The use of density transfers is a more controversial tool, partly because of their complexity. The community, through its community plan, designates areas where the development potential may be transferred from one "sending" property to another "receiving" property. Rezoning of each of the properties is then required to enshrine the appropriate density. The sending property, having environmental values, is protected while the receiving property receives a greater concentration of development. The recipient of the density pays the owner of the protected land and the negotiation is carried out without involvement of the local government.

Local bylaws can control disposal of storm water or surface runoff from private property (from roofs or areas to be paved). These bylaws specify disposal location, maximum area for paving with impermeable material, and a requirement for water purification devices; other tools are sampling for contaminants to pinpoint the need for repairs, an atlas to show the presence of storm water discharges, and model storm sewer and watercourse protection bylaws (Capital Regional District Environmental Services 2007). These conditions augment efforts made at time of subdivision, or in large developments, to build retention systems, filtration systems, and bio-retention or constructed wetlands to aid in removal of pollution.[57]

Also at a local government level, green roofs may be encouraged; vegetation replaces gravel and other common commercial roofing materials. Benefits include better storm water management (one study reports a 54 percent reduction in runoff (Liu and Baskaran 2005)). Taking stormwater management to a watershed scale, "rainwater management" would restore the function of trees, soil and open space for natural absorption, storage, evaporation and filtration services, rather than relying on pipes and culverts (University of Victoria Environmental Law Clinic 2010, 8).

Besides stormwater, pollution from septic systems, fertilizers, underground storage tanks and certain commercial establishments requires control. Septic system regulations specify proper siting, design and construction as well as a permitted density based on the assimilative capacity of soils and watercourses. Underground oil storage tanks may leak if they become corroded or are improperly used or maintained, while leaks and spills may occur from above ground storage tanks. Clean-up of contaminated areas and replacement of older tanks that may corrode and leak oil can be required. Prince Edward Island enacted such regulations given their dependence on groundwater. Their program is to be phased in, but ultimately will require licensed installers and inspections, denial of fuel oil delivery if an inspection has not been carried out, and tank replacement every 15 to 25 years (Prince Edward Island Fisheries, Aquaculture and Environment c2001).

Finally, sewer use bylaws deal with certain operations: auto repairs, carpet cleaning, dental operations, dry cleaning, fermentation, food services, photo imaging, printing, and vehicle washing. For example, dry cleaning operations must not discharge prohibited or restricted wastes and wastewater containing tetrachloroethylene in concentrations greater than specified, tetrachloroethylene-contaminated residue; or uncontaminated water in quantities greater than specified volumes. Fermentation operators must test any wastewater containing acid or caustic cleaners or sanitizers for pH and adjust the pH to specified limits prior to waste discharge. Liquid waste from vehicle washing businesses must be equipped with vehicle wash interceptors to treat the waste discharge (Capital Regional District 1994).

Community-based Projects

There are so many wonderful examples of community-based volunteer efforts to protect the aquatic ecosystem and these often benefit from government funding and leadership. Examples include south-western Nova Scotia's Clean Annapolis River Project (2008) and the joint project between Québec and the Government of Canada, the St. Lawrence Vision 2000 Action Plan (2003). These projects have raised awareness, collected information and completed mapping, improved water quality, undertaken restoration and conservation projects, and established monitoring programs.

Also on a community level were 43 Remedial Action Plans prepared under the *Great Lakes Water Quality Agreement*. In the Collingwood Harbour area, based on a plan integrating land and aquatic ecosystem components, improvements enhanced the municipal sewage-treatment plant, protected an existing wetland complex, rehabilitated fish and wildlife habitat, and educated the community (Mitchell 2005, 1345). The Hamilton Harbour area of Lake Ontario illustrates the generation of social capital: 1000 individuals from clubs, conservation authorities and local industry participated in a Remedial Action Plan process to develop *Vision 2020: The Sustainable Region*. Despite these promising efforts and others, only three areas are delisted and disputes remain between the many players involved at government, industrial, port, recreational, environmental and other levels (Sproule-Jones 2008, 201).

Best Management Practices

Best management practices encourage appropriate behaviour where activities may deleteriously affect the aquatic ecosystem. Usually issued in the form of guidelines, best management practices only sometimes become part of formal approvals and legislation. For farming, such practices include fencing to keep livestock away from streams, hardened access points for animal crossing, planting crops that will use excess soil nutrients and hold soil in place, and using conservation tillage. Waste management guidelines direct manure application and disposal of manure, milking parlour wastes, wood waste and dead animals (British Columbia Environment, Lands and Parks. 2001). These practices and others such as rotational grazing, mechanical harvesting, small reservoir retention and controlled tile outlets were evaluated as part of the *Greencover Canada* project, hopefully leading the way to new methods of protection (Theriault 2006, 6). Best management practices rely for their effectiveness on education and training, discussed in the next section.

Education and Training

Education and training are successful tools in water management: augmenting a regulated activity, and where regulations do not apply, encouraging appropriate behaviour. The federal government's *Pleasure Craft Sewage Pollution Prevention Regulations* under the *Canada Shipping Act* only apply in designated areas; for these and other areas, brochures alert all boaters to actions they can take (BC Environmental Protection Division 2001). Similarly, there is an education program to prevent consumers from sending batteries to garbage dumps (Canada Environment 2007). Education can also be the primary work of a volunteer group. "Cows and Fish" is another name for the Alberta Riparian Habitat Management Program. They teach farmers how to use responsible grazing and land management practices along the banks of streams, rivers, lakes, ponds and wetlands (Brinkman 2003, 47).

The 8 Canadian groups who are members of the worldwide Waterkeepers Alliance combine stewardship, advocacy for stronger laws or community support, and education to defend their local waters. Stream of Dreams Murals Society sponsors the wire fences around schools decorated with brightly-coloured fish in wavy continuous streams. This organization's mission is to "educate communities about the life and function of their watersheds, rivers and streams, while dazzling them with the charm of community art," which they surely do (Stream of Dreams 2011). As of 2011, they have 125,000 participants in three provinces.

This section gives a lengthy description of the means available at various government and non-government levels to protect the aquatic ecosystem—much is achieved, but active coordination of all efforts is needed; so too is protection of water for its own sake.

Honouring the Intangible Values of Water

Put simply, the issue is who will speak for the stream? In a more complex explanation, how can we "appeal to ethical and aesthetic principles, rather than economic calculation" (Norton 2006, 337)? How can we take into account intrinsic values of water?[58] How can we show that there are times when the highest value use is no use and that protection from harm must occur at all costs? How can we best achieve protection of special water bodies? There are two aspects: how to express intrinsic values (to date, mainly the role of economists); and second, how to effectively insert arguments for these values into decision-making processes. Currently, these activities occur during public participation efforts and planning processes, but consideration needs to be commonplace in everyday decisions and cope with differing individual valuations. This is difficult when issues are complex and economic values are more readily available, but an ethical approach to water management requires recognition of aesthetic, spiritual and environmental preservation values.

Exploring works on the role of water in language and culture, in landscape design or ecopsychology, and on water's symbolism in mythology and religion could be useful, but these are not discussed here. Five approaches that begin to right the balance come from conservationists, landscape aesthetics, new valuation methods, historical geographers, and from art, poetry, literature and musicians. These methods, while primarily used by government officials and academics, could be adapted by groups and individuals seeking to protect water's intrinsic values.

It is obviously not sufficient to rely only on government-created parks, protected areas for endangered species, and water bodies conserved as special places, but there is much to learn from these situations. Supported by the federal, provincial, and territorial governments, the Canadian Heritage Rivers Board, with local support, promotes, protects and enhances Canada's river heritage. Rivers in the scheme use a classification system designed to reflect heritage goals (Canadian Heritage Rivers Network 2001). Unfortunately such preservation and conservation efforts are not always possible. We need other tools to make decisions in the context of proposed development.

The methods of landscape aesthetics have evolved since the 1960s from quantification, to expert appraisals, to user preference models; a veritable smorgasbord with no apparent consensus (Porteous 1996, 193-208). Their use to express values associated with water environments is nevertheless recommended, particularly where augmented by participant perspectives (Dakin 2003). Historical geographers' work as represented by a 2007 conference entitled *Confluences: A Workshop of Rivers, History and Memory* should also be consulted. The presentations show the relationship between rivers and public history and memory (NiCHE 2007).

Beyond this, water managers need other ways to incorporate the value of water for non-consumptive uses into their decision-making. Inferential valuation (contingent valuation) uses data of actual purchases of goods and services to infer the value of a non-market resource; for example, looking at the money and time costs incurred to enjoy a visit to a lake or asking a sample population hypothetical questions about their willingness to pay for retaining or restoring a wetland (Kahneman 1992, 57-60; Lantz et al. 2010). Hedonic pricing looks at the value of market goods such as real estate to determine the difference between, for example, lakefront and other property. This information can then be used by policy makers who evaluate the benefits of preserving water environments.

Where there are complex choices to be made, multi-attribute analysis provides a decision framework that integrates values and technical information and, where development is proposed, answers the question: "Is it worth it?" "[G]iven a context of social diversity, it deals with the question of the loss to our society and our region for a given decision and informs the nature of compensation for environmental damage" (McDaniels 1993, 21). This approach identifies "objectives", determines alternatives or options in relation to these objectives, analyses and ranks the impacts of alternatives, clarifies trade-offs using currency or other values, and then evaluates the alternatives. All of these valuation approaches have a dollar value in mind, the latter perhaps less so, suggesting the need to look elsewhere.

The emotionally imaginative work of painters, poets, writers and musicians often portrays the intangible values of water. As has been said in another context, these values are our ""second nature" that typifies the way a local community or culture understood its relationship to earth …" (Kingwell 2008, 34). Sufficient place needs to be given to these sources of inspiration—to welcome them and to learn from them. Although creative sources may be considered elitist by some, focusing on local contributions in combination with oral histories and historical accounts, can enunciate the intangible values of water.

There are many examples of honouring Canada's water, even when we look just at literature and art. *Thinking about Water* explores water's cultural meanings and ecological issues (Chen et al. 2013). Environment Canada publishes a discussion online about Canadian artists and their depiction of water and about water and Canadian identity (Canada Environment 2004d). Joan Murray's book *Water: Lawren Harris and the Group of Seven* (2004) describes an image inspired by Canada's water: "The sketch of Harris's *Agawa Waterfall* is a tribute to the ideal of water. It shows us the typical Harris space, which involves water as an interweaving of prospect and refuge. The form of the water, which seems to stream towards us in the foreground, and the rocks and distant trees in the background, give dynamic but unified shape to the tension caused by the waterfall itself." Similar analysis relating to the work of local artists would assist; for example, the wonderful photographs of Ontario lakes from fine art photographer Janusz Wrobel.

A search of fiction, non-fiction and poetry also reveals many examples. My water management class for the University of Victoria's Geography Department experienced this in the final lecture. While viewing slides of work by members of the Ontario Society of Artists, representing their concern about the quality and safety of drinking water in the aftermath of Walkerton[59], each student read aloud a work of their choice. Reviewing 500 examples, the breadth of their selection is amazing. The poem I chose to read was always the same, Al Purdy's last revised poem

(Purdy 2000, 579) in which he exhorts us to *Say the Names*; many are watercourses (for example, Tulameen and Similkameen)—looking back to my first encounter with water management, advice to know the names of all the watercourses made even more sense.

Future Actions

Protecting the aquatic ecosystem is certainly a theme whose time has come—witness environmental education for children; see the fish painted on pavements near drains, often with the message "Dump No Waste, Drain to Stream", to know that the next generation will have firm views on this subject. And they are the best educators, bringing the message home. The bad news is that as we develop protection measures, land development adjacent to fresh water sources is seen as more acceptable. This suggests the need to err on the side of caution[60] when developing prescriptions for ecosystem protection.

While allocating water, decision-makers should take into account both equity and efficiency considerations. Achieving efficiency suggests that water be allocated among competing uses to provide the greatest benefit to society. Counting benefits based on a dollar value is not sufficient. Increasingly, stream channels and flows for fish, wildlife and the environment need to be preserved in the public interest. A growing number of jurisdictions recognize environmental values in principle, but progress is slow and other problems persist. In particular, these problems occur in protecting water's quantity and quality; for example, not addressing linkages between groundwater and surface water (Annear et al. 2009, 81). Adequate protection from non-point sources of pollution such as agricultural operations may be lacking; Montpetit (2003, 90-104) found that the Ontario agricultural sector's desire to maintain the status quo mitigates against change.

Policy-makers charged with developing guidelines and education need to consider whether legislation would be a better approach than reliance on guidelines. In addition, there is growing interest in environmental markets, some of which have already been described. These include compensating pollution sources for reducing emissions, income for agricultural producers who reduce high-nutrient runoff, water trading, conservation easements for habitat protection and habitat banking (Sustainable Prosperity 2012, 8).

As to the intangible values of water, developments in non-market valuation, landscape aesthetics, and historical geography are important. What we also need is a Canadian consciousness around water that goes well beyond the Heritage Rivers' scheme. When local governments develop sustainability checklists to judge land use applications, preserving the intangible values of water should be there. "Waterspaces (view and recreation sites, transportation links, ecosystem entities, archaeological and historical sources, and the presence of nature)" (Pinch and Munt 2002) remind us of the interconnections in our lives and in the life of a river.

Conclusion

To answer the question of who speaks for the fish and who for the stream, this chapter showcases responses from government and non-government perspectives; there is still much to do. We need "to illuminate the link between functional riverine resources and quality of life issues" and to increase knowledge about water's availability and allocation (Annear 2009, 71). Provincial governments have slowly adopted protection measures; an increase in low flow situations suggests the need for more immediate action. Local governments and volunteer groups take the lead. More comprehensive planning and integrated resource management schemes would reap the benefit of these activities (see Chapter 11); for example, augmenting volunteer efforts relating to data collection and remediation. Also necessary are innovative education efforts, recognition of groundwater-surface water connections, increased stormwater management, translation of best management practices into regulation (where guidelines and education are just not sufficient), and monitoring efforts.

Translating the intangible values of water into a format usable by decision-makers is difficult. The emotionally imaginative work of poets, painters, artists, and musicians, along with people's stories and the work of water historians, bring a voice to the ecosystem. Many years ago, Sagoff (1974) wrote that moral and aesthetic sensibilities guide protection of landscape, species, waterways, and marshes, but accomplishment rests on political compromise at the local level. Honouring intangible values needs to be embedded in our environmental awareness and activities.

CHAPTER 7: PROTECTING DRINKING WATER QUALITY

... till taught by pain
Men really know not what good water's worth; ...
(Byron. Don Juan: Canto the Second, stanza 84).

It hardly seems possible, ten years since the Walkerton tragedy. In a news release, Ecojustice Canada reveals the existence of 1,776 drinking water advisories in Canada, 116 First Nations' communities under Drinking Water Advisories for risk of waterborne contaminants, and 20 to 40 percent of all rural wells having unacceptable coliform or nitrate concentrations. Less than half of Canadian provinces and territories require "advanced" treatment of surface water, a standard practice in the European Union and the United States (Ecojustice 2010).

News releases catch our attention. The World Wide Web confirms the importance of this subject. Canadian water quality[61] is a striking example. Using *Google Canada* in May 2003 to search for a combination of "water quality" and "Canada" produced 655,000 returns. The search tool formula may have changed, but in January 2013 there are 6,670,000 returns. Many relate to government activities, not least of which are assurances as to drinking water's safety. Given the plethora of sources, this chapter briefly summarizes information about the cause and effect of problems. It then describes activities to protect drinking water quality and the essential role for government, source water protection planning, and non-government actors.

Problems Defined

Untreated water can have many problems. Natural elements cause drinking water to be unsafe: "On the Prairies ... arsenic in groundwater is locally above safe limits; in the Moncton area, fluoride levels are high; and in large parts of the Northwest Territories, elevated levels of radioactivity are observed" (Canada Environment 2003). Soils eroded through natural or human-induced events cause turbidity. This affects the ability of disinfectants to eradicate pathogens before they enter a water distribution system. In November 2006, heavy rain lasting for 15 hours hit the Greater Vancouver Water District watersheds; turbidity from mudslides clouded the water supply making it unsafe for drinking, brushing teeth, or washing food (Mickleburgh 2006, S1). There is always a chance that the water is contaminated by animal feces and is carrying *Giardia*, *Cryptosporidium* and *Escherichia (E.) coli* bacteria. For residents, this meant 12 days of boiling their water.

Watershed diagrams, such as the one used in 1999 by BC's Auditor General in a critique of water quality protection, did not foresee the possibility of mismanagement of drinking water provision as a contamination source. Nor did officials foresee what a high profile issue this would be—witness the events at Walkerton and North Battleford (see Hrudey and Hrudey 2004). Even with every effort to protect drinking water, disease outbreaks occur. Frequently, their causes are an agent such as pathogenic bacteria (for example, *E. coli* O157:H7), viruses, parasites and other organisms. Parasites such as *Giardia* or *Cryptosporidium* resist disinfection through chlorination, ozonation, or removal through filtration.

And watershed diagrams infrequently show two other problems. First is a common event, water pollution caused by flooding. In October 2003 some residents of Squamish, BC had to evacuate their homes after an exceptionally heavy rain. Flooding damage was only part of their misery. Returning, they found that their water was suspect and subject to a boil water advisory.[62] Extreme storm events, often attributed to climate change, play a major role in these situations. Lead is another problem; naturally-occurring in bedrock, it can dissolve into water supplies, but also can result from the pipes or soldering of pipes in homes built prior to 1955. Nine percent of homes in Toronto have service pipes made of lead (Howlett 2007, A9).

Various risks to health from drinking contaminated or polluted water include microbial agents and indicator organisms; viral agents; and parasites such as *Giardia*, *Cryptosporidium Toxoplasma Gondii* and *Cyclospora cayetanesis*. Physical and chemical parameters include turbidity and arsenic, nitrites and nitrates. Persistent organochlorines include dioxins, furans, and DDT. Also of concern are radionuclides in uranium and radium from soil or rocks; disinfection byproducts and unpleasant aesthetic parameters and odours (BC Health Planning 2001). One example that has received considerable media attention is the radionuclide tritium, a contaminant in water supplies, with possible damaging levels coming from leaks at nuclear power plants (for example, Pickering, Ontario) and from sign companies. An Ontario manufacturer of glow-in-the-dark signs lost its licence after tritium levels as much as 743 times the normal were found (Canadian Broadcasting Corporation 2007b; Mittelstaedt 2005, A5).

In addition, there are the effects of fluoride and pollution from harmful doses of drugs (antidepressants, antibiotics, cancer drugs and drugs used in farming) released into water. Some fear terrorists' release of toxins such as ricin into water (Bronskill 2003). Fluoride, while lauded for causing a decline in tooth decay, has another side. It may increase the risk of fractures, degenerative bone diseases, and dental fluorosis (pitting of teeth). Other suggested side-effects are: a higher risk of Down's syndrome, a link to osteosarcoma, reduced intelligence in children, and impaired thyroid function (Mittelstaedt 2007, L1). When drinking water problems occur, the effects are far-reaching.

Effect of Drinking Water Quality Problems

A *Google* search reflects the value people attach to good drinking water quality. And the reason is clear. Water-related health hazards include acute problems such as nausea, vomiting and diarrhoea. These may be short term, others can be debilitating and in extreme circumstances, lead to death. How frequently do these problems occur? Thousands of Canadians suffer from gastrointestinal illness every year. Although it is not yet known how many of these come from waterborne sources, a link is evident between illness and warm conditions over a 6 week period, particularly in Alberta and Ontario, and to extreme precipitation (Waltner-Toews 2005).

The effect of water quality problems is devastating and may go well beyond health hazards. Joy Parr, then Canada Research Chair, Technology, Culture and Risk, in the Faculty of Information and Media Studies at the University of Western Ontario, paints a clear picture:

> The contaminated water townspeople drank changed bodily experience of being in Walkerton. These were days lived amidst the merged sensations of pain in the gut and the mingled smells of chlorine and excrement by the many local sickbeds in town, and overhead too many times there could be heard the dark beat of helicopter blades as the most gravely ill were evacuated to the nearest university hospital. In town all these experiences were understood as implications of the long-standing compact the community had made about their water and its guardians (Parr 2004, 257).

So much has been written about the Walkerton tragedy in newspapers, journals and books (for example, Burke's *Don't Drink the Water: The Walkerton Tragedy* (2001) and Perkel's *Well of Lies* (2002). This is as it should be in response to a tragedy caused by *E. coli* O157:H7 and *Campylobacter jejuni* bacteria from manure spread on a farm near the infamous Well 5 in Walkerton. The Province of Ontario has paid $72 million compensation to the victims since 2000 (Canadian Press 2011). Proper practices were followed on the farm. The reasons for the outbreak are more complicated: untrained waterworks staff, falsification of records, false information about contamination at the outbreak, insufficient provision of legal standards, and heavy rains that washed manure into wells. Cuts to the Ontario Ministry of Environment budget led to the privatization of water testing and inadequate inspection of municipal water systems by provincial officials. While in 2006, 85 percent of the town residents indicated that they are in good to excellent health, some people still suffered from high blood pressure and kidney problems (Globe and Mail 2006b, A13).

Walkerton was followed, in April 2001, by a *Cryptosporidium parvum* parasitic contamination in the Battlefords area of Saskatchewan that caused gastrointestinal illness to up to 7,100 people. Here investigators found a malfunctioning filtration system, staff shortages, rodent droppings in the well, and untreated sewage released in the river upstream from the water treatment plant. Robert Laing, a commissioner for a subsequent inquiry, said: "there was simply a great deal of indifference to the public health safety aspects of drinking water on the part of the city, who had the responsibility to produce potable water and on the part of SERM (Saskatchewan Environment and Resource Management) who had the mandate to regulate it" (*Regina Leader Post* 2002). In 2003 an out-of court settlement amounting to $425,000 was reached with 100 people who became ill in North Battleford; previously there had been a $3.2 million settlement with 700 people (*Globe and Mail* 2003, A7).

Justice O'Connor in his Part 2 report on the Walkerton tragedy recognizes that drinking water infrastructure for Métis and non-status Indian communities and First Nations reserves is obsolete, entirely absent, inappropriate, or of low quality (O'Connor 2002, 486). For example, the Kashechewan Reserve, 400 miles north of Timmins, Ontario was subject to a boil water advisory for two years prior to a water sample in October 2005 that showed the presence of *E. coli*. The event led to evacuation of 100 people and a call to relocate the community. The cause of the problem: at least one water intake pipe is located downstream from where raw sewage from the community flows into the river. To add to this, the community floods each spring (Canadian Broadcasting Corporation 2006a; Priest 2005, A5; Strauss 2005a, A4). Management of the community's water supply is now vastly improved, but drinking water quality problems have precipitated action at all government levels as the following sections show.

Government Action

Federal responsibilities are divided between three departments. Environment Canada and Health Canada (*Canadian Environmental Protection Act* (1988, renewed 1999)) assess and manage threats posed by toxic substances. Health Canada (*Food and Drug Act* and *Guidelines for Canadian Drinking Water Quality*) in partnership with provinces and territories ensure that bottled water containing poisonous or harmful substances is not sold in Canada. Environment Canada has responsibility to prosecute against offences to environmental laws, assess and control the most dangerous chemicals, provide knowledge (Canadian Water Network) and environmental impact assessments, and encourage technology. It also funds community projects that support clean water. Health Canada and Aboriginal Affairs and Northern Development Canada (with Band Councils, Environment Canada, provincial governments, and sometimes with municipalities) fund water services on reserve lands. The Federal government is also responsible for water on federal lands, in national parks, on national and international transport, and for providing safe water to federal employees.

A specific example is the division of responsibility for First Nations communities: Indian and Northern Affairs funds the capital cost of plants and piped systems and a portion of their operating and maintenance costs, enforces certain standards through funding agreements, and provides engineering advice and approvals. Public Works and Government Services Canada assists with procurement. Health Canada delivers drinking water monitoring programs on reserves located south of the 60th parallel, either directly or in an oversight role. Environment Canada promotes source water protection and enforces effluent discharge standards.

Funding to protect drinking water quality for these communities for 2002-2007 was $16 million (Swain et al. 2006, 20, 23). In 2011, the Auditor-General reports that more than half of the drinking water systems on reserves still pose a risk. Forty percent of the 1,180 First Nations' homes that do not have water are located in 4 communities in the Island Lake region of Manitoba (Galloway 2011, A8). Progress is slow with lack of trained personnel, funding issues, and narrow construction windows. Appropriate monitoring and testing is being carried out, but only 25 of 80 required annual inspections and 47 of 80 risk evaluations in the Auditor General's sample were carried out between 2006 and 2010 (Canada Auditor General 2011).

Most jurisdictional control relating to drinking water quality is provincial, territorial or local. Protective actions have increased dramatically and include information management, guidelines, liquid waste management, a multi-barrier approach to the protection of drinking water, legislating procedures and standards as well as monitoring these standards, planning to protect source water, and seeking funding for infrastructure.

Information Collection and Management

Baseline information on the potential for contamination from present and future land uses and of ambient (immediate surroundings) water quality is essential to manage quality and mitigate problems. Government agencies rate individual water bodies using a Water Quality Index; assigning a numerical order shows areas for improvement. "State of the Environment" reporting is a recipient of this useful information.

Water Quality Guidelines

In decision-making, most jurisdictions use water quality guidelines, benchmarks for water quality for various uses. *Guidelines for Canadian Drinking Water Quality* have existed since 1968, a product of the Federal-Provincial-Territorial Committee on Drinking Water in collaboration with the Canadian Council Ministers of Environment. Guidelines give "safe limits" for various polluting substances in raw (untreated) drinking water or for other uses of water (Canada Environment 2004e).[63] Ambient water quality guidelines are a provincial tool to allow a water manager to take into account the natural quality of the water body and threats from local sources when dealing with drinking water and aquatic ecosystem protection, as well as other water uses. Often these objectives have no legal standing; for example, in granting a water licence. Outside of territories subject to the Northern Authorities' Management Scheme and Québec (see Chapter 3), water managers are left to "best judgment" when issuing water licences or permits. Should guidelines become law? Based on the experience at Walkerton, Ontario's Nutrient Management Act (2002) is a good example of such a change as it seeks to minimize the effects of livestock manure and other nutrients stored on farm properties or applied to land.

Liquid Waste Management

Preventing wastes from entering water bodies in the first place, treatment of industrial wastewater and sewage, and stormwater treatment are essential steps in protecting water quality.[64] But liquid waste management is too big a subject for this book. Detailed information is available from the Canadian government's National Pollutant Release Inventory. Readers may wish to consult Benedickson's *The Culture of Flushing: A Social and Legal History of Sewage* (2007) and *The Walkerton Inquiry Commissioned Paper 1—Water Supply and Sewage Infrastructure in Ontario, 1880-1990s: Legal and Institutional Aspects of Public Health and Environmental History* (2002).

Multi-Barrier Approach to Drinking Water Protection

Treatment for drinking water quality has a long history culminating today in the "multi-barrier approach". Johnson (1988, 136-139) gives a useful overview, divided into three parts: ancient, when individuals controlled water quality by keeping water in copper vessels, exposing it to sunlight, and filtering it through charcoal; progressive, beginning in 1880 and seeing the eventual development of standards, filtration methods and chlorination; and contraindicative, recent concerns about water contamination and lifetime or chronic health effects from water treatment and substances of unknown significance.

For large water systems and even for small communal systems, where resources are available, the tools and procedures of an integrated multi-barrier approach are available.[65] There are five steps, the first four of which reduce or eliminate possible pathogens and contaminants. These include protection of water sources (discussed in a later section), water treatment (disinfection and/or filtration by the purveyor, guided by a health authority), maintenance of clean storage and distribution systems (reservoirs, stand pipes, water mains, valves, hydrants) by a purveyor guided by a health authority and with local governments, and comprehensive testing/monitoring and enforcement of standards by a health authority in partnership with the purveyor. If problems arise, then a fifth step is essential; response must occur in a well-thought-out, thorough way (O'Connor 2002, 74). The Canadian Council Ministers of Environment (2004) stress the importance of this risk management approach. Additional benefits noted are more stakeholder (customer) involvement and understanding, clear information for politicians who must supply leadership and funding, better source protection, better education of staff who operate water systems, and good preparation for emergencies.

In response to concerns about drinking water quality, the technological interventions of treatment are many (British Columbia Health Planning 2001; Canadian Broadcasting Corporation 2006c).[66] Examples include the use of filters, coagulation, flocculation, sedimentation, and chlorination.[67] Newer forms of treatment are ozone oxidation and ultraviolet light, although these are ineffective outside the treatment plant. Running an electric current through air or oxygen creates ozone. The resulting ozone bubbles through water to kill microbes. Ultraviolet radiation disinfects water by shining ultraviolet light through water killing parasites such as *Giardia* and *Cryptosporidium*, but is most effective in low turbidity water. Finally, secondary disinfection uses chlorine at points of the water distribution system. In addition to these purification systems, point-of-use water filters are available to attach to home water systems or at water taps.

There are elaborate schemes to safeguard water in large urban areas; the Halifax Regional Water Commission uses a multi-barrier approach. It designates watersheds for the two supply sources, Pockwock Lake and Lake Major, as protected areas under the *Nova Scotia Environment Act*. Commission and provincial staff and the community restrict and monitor activities within the watershed. Two large and seven small water filtration treatment plants use colour precipitation (clarification, filtration, disinfection and corrosion control) to ensure adequate water quality. The Commission rehabilitates about 5 kilometres of water main each year to repair leaks and protect water quality; cross connection control and backflow prevention minimize pollution. Continuous monitoring and testing ensure that that water quality guidelines are met (Halifax Regional Water Commission 2006). The Halifax Regional Water Commission has every opportunity to provide water quality protection. For operators of small water systems, it is more difficult.

When faced with quality problems in the water supply, one local water official said: "I was baby-sitting on a 24-7 basis making sure there was enough water to fill up reserves. I had to put on water restrictions and I'm only a one-man show here. I have to go out and police water restrictions. Sometimes I had to put on a blackout where I turned off the water until the reserves filled up and I was getting a lot of calls from those people saying 'we're out of water'" (Theodore 2003). For small operators, challenges may include "lack of awareness and training about proper operation and monitoring, and lack of financial resources to safeguard drinking water. In some cases, systems are abandoned or there is no designated purveyor" (BC Drinking Water Review Panel 2002, 472). Further, more elaborate treatment may not make economic sense for these operators (Swain 2009, 27).

Legislating to Protect Drinking Water Supplies

Walkerton was a crisis that led to significant legislation and strategy relating to drinking water quality across Canada. Review of legislation from four provinces (British Columbia, Alberta, Manitoba and Ontario) finds a number of common provisions. These include provincial oversight: monitoring and enforcement powers, review panels and general standards and provision for planning. Municipal accountability is in the standard of care for oversight of systems. Waterworks operators have performance and testing standards. Laboratories may be required to report non-compliance with standards and any health risks to suppliers and the public may be given access to test results, approvals and orders. Consumers are to receive an annual report about their water quality.

Despite its length, the list is not exhaustive and every province has different legislation, a reaction to the many kinds of water suppliers. Government authority, while necessary given the seriousness of this public health issue, tends to exclude the involvement of others. Provisions for source water protection planning are more recent additions in a number of provinces (e.g., Saskatchewan and Ontario) and may rectify this situation.

In a study of progress to date, the Sierra Legal Defence Fund (Ecojustice Canada)'s *Waterproof 2: Canada's Drinking Water Report Card* criticizes the federal government for problems on First Nations' reserves, lack of leadership and a hands-off approach relating to water quality, slowness in developing guidelines, and ignoring climate change (Christensen 2006, 30-33). The Report Card (2006, 6, 42-45) sees improvements by provincial authorities since 2001 in water treatment, binding contaminant limits, testing, operator certification and public transparency, but there is still insufficient action. By November 2011 and Waterproof 3, there is not much improvement (Christensen 2011). The federal government receives a failing grade, but at least Ontario (A) and Nova Scotia (A-) earn kudos for their work on drinking water protection.

In particular, we need more funding to upgrade and maintain small community systems. Local governments in Canada consistently lobby senior governments for provision of adequate infrastructure grants for water and other large projects. This is not surprising when you consider the costs; replacement cost for drinking-water infrastructure in "fair" to "very poor" condition is $25.9 billion and for wastewater infrastructure in like condition is $39 billion (Federation of Canadian Municipalities 2012, 3). Local governments must seek creative funding, including private-public partnerships. Other options are an increase in user fees or searching for private capital investment. Incentives to maintain water quality can also include fees for point-source polluters, large penalties for environmental damage, taxes or subsidies on receiving water quality, and ensuring that polluters assume liability for water quality damage (Coote and Gregorich (Eds.) 2000, 109).

Planning for Source Water Protection

Protection of source water quality for drinking water is conceptually and practically a difficult step. Ultimately a form of land use management, it involves many players: the owner of the water system, owners of land around the water source, the public, users of other resources in the watershed, recreational users, and relevant government agencies (Ontario Environment 2003). A senior engineer interviewed as part of a political ecology[68] study of BC's Okanagan Valley said: "If you ever phoned (the BC Government) to find out who is in charge of your watershed you'd get the switchboard from hell. I think it is authority that is the key word we are looking for. Who has the authority in the watershed to deal with drinking water issues? In my opinion they are not there." A professional engineer working as a consultant added "We've got a great number of agencies that are managing their own silos, if you will, and a whole series of silos that are not talking to each other very well, and really no one talking for water. A whole bunch of little voices talking for water, but no one voice saying I am water and I am responsible for what goes on here" (Patrick 2009, 213).

Their comments come as part of a study of who has the power to make change in source water protection. It finds that change tends to concentrate at the local level (multi-purveyor joint water committees, different watershed user groups, education and information efforts, and water purveyor staff) (Patrick 2007). In contrast, the provincial level, in this case with authority for source water protection, restricts change. Patrick suggests that many different agencies and inter-jurisdictional rivalry exist; de-regulation and re-regulation lead to local-level confusion. As to the case-study area, perhaps these issues will be resolved by the addition of a water management initiative to the Okanagan Basin Water Board and their future planning.

Despite the costs of such planning exercises, the benefits are significant: reducing risk in a cost-effective manner and keeping contaminants out of drinking water in the first place (British Columbia Health Planning 2001, 80; Lindgren 2003, 11). As Justice O'Connor reports: "The public strongly favours source protection as a key component of our water system. No other aspect of the task of ensuring drinking water safety received as much attention during the town hall meetings that this Inquiry held across Ontario" (O'Connor 2002, 8-9).

Non-Government Activities

While there are many non-government activities, including those described in Chapter 6, two examples are efforts by golf course managers to change their method of turf production to protect both drinking water quality and the ecosystem and water system operators to update their skills. Golf courses can join the Audubon Cooperative Sanctuary Program that helps "people enhance the valuable natural areas and wildlife habitats that golf courses provide, improve efficiency, and minimize potentially harmful impacts of golf operations" (Audubon International 2007). Training for water-system operators is a focus across Canada, a special example being the Walkerton Clean Water Centre. Courses offered by organizations like the BC Water and Waste Association include many subjects: water treatment, wastewater collection and treatment, operation of small water and wastewater systems, chlorination, cross connection control and water quality for distribution operators.[69]

Future Actions and Conclusion

Since 2000, activity associated with drinking water quality has been frenetic. Yet problems persist, particularly for aboriginal communities. It seems reasonable that time is needed to allow reflection, monitoring and continued improvement. But to achieve consistently high standards of drinking water quality across Canada, there are opportunities for improvement: adequate funding for building and maintenance of water supply and water treatment systems, investment in technology and data, clear lines of authority, well-trained operators and better communication. Christensen (2003) calls for national standards for drinking water quality testing and treatment and an expanded list of contaminants[70]; we need to know the characteristics of all water systems, water system performance, and the occurrence of water-related disease, as well as the results of risk assessments and evaluations (BC Health Planning 2001, 3; BC Ombudsman 2008). In addition, water managers need better ways to communicate the risks in order to influence policy-making; academics to find effective ways to let science speak to policy (Davies and Mazumder 2003, 273).

One of the most apparent issues in this chapter is the number of different groups involved in water quality management. As a public safety issue, government involvement is essential, but finding both vision and leadership and determining the role of local government and non-government agencies is key to future success (de Loë and Kreutzwiser 2003). Better integration between governments is essential; flaws exist in linkages between responsible agencies (Lebel 2008). This integration needs to be more pervasive and involve the local level so that all types of planning incorporate water treatment and source water protection and control of urban land and rural resource use. Finally, the uncertainties presented by population growth and climate change loom over all the other challenges (Fraser Basin Council 2005a, 18), and, add to these, restrained economic times.

This chapter's emphasis on drinking water quality protection responds to the place of human health on the Canadian public's agenda. We need institutional change and coordination between agencies; strong leadership at political, bureaucratic and interest group levels; and greater funding. Also needed: more data, more training for water system operators, opportunities for public participation, greater transparency in decision-making, and finally, an enhanced focus on the effects of development on water quality—not a small order! These suggestions are not new and will continue to be raised. We learn the hard way; there is still more work to do.

CHAPTER 8: FLOODING: NOT A QUESTION OF IF, BUT WHEN

As the snows melted and the brown water inched higher, at first their dike held. But eventually, the huge weight forced the wall to buckle. The water enveloped the first floor and was lapping at the upstairs windows before it crested. "Of course, everybody reassured me that it could never happen," said Mr. Bloomfield, who lives in a riverside home in the Winnipeg suburb of Saint Norbert. "But the dike collapsed, the windows broke and water just flooded in" (Strauss 2005b, A3).

Protection from flooding is the third topic in the theme "public health and safety and protecting the aquatic ecosystem". Whether you are a flood victim or a general taxpayer, you are affected by flood damages. Perhaps the only people who show a positive attitude to flooding are those who work in "floodplain management", a term used to describe actions taken to protect people and property against flooding events. More recently, it has also come to mean avoiding damage to ecosystems in the floodplain.

This chapter examines flood events to show why action is necessary and explains what is done to prevent flood damage and alleviate personal suffering and the great cost when floods come. These interventions typify the way in which water management has changed from a "command and control", often single-purpose, activity to one involving local level planning and integrated decision-making. The chapter illustrates the importance of gaining political acceptance for proposed activities, how the media bring flood events to public attention, and the need for public education and involvement. It explores the historical context for flood protection, using British Columbia's Lower Fraser Valley to illustrate the concepts—a success story in many ways, but also the epitome of a "wait and see" attitude, of which we are all guilty.

Experiencing a Flood

If you live on the floodplain of the Bow and Elbow Rivers, particularly in Calgary, Canmore, Medicine Hat, High River (the hardest hit) or the Sisika First Nation, your experience of flooding is both recent and traumatic. From June 19 to 21, 2013, up to 300 millimetres of rain fell adding to rivers that were already full from the Rockies' spring runoff. In the resulting flood flows, three people died, property and infrastructure damage may exceed $5 to 6 billion, and power outages closed Calgary's business district. Newspapers, television, the internet and social media gave daily coverage and showed the images of devastated communities. In Calgary, 100,000 residents from all income levels had to leave their homes. At the height of the flood in High River, the town was off-limits for its 12,900 residents with the first able to return after ten days, but some not for weeks. They were evacuated by any means available including combines and front-end loaders. Their evacuation centre and hospital's first floor were flooded.

As the waters receded, 2,500 volunteers in Calgary turned up at the first call. Mayor Naheed Nenshi and Alberta's Premier Alison Redford were praised for their leadership. The flood-damaged Calgary Stampede site was prepared for its many visitors in just 9 days. The Calgary Zoo was to re-open slowly as repairs are made. Using social media, many offered accommodation to displaced residents. Affected residents received pre-loaded debit cards for immediate recovery costs. The Red Cross received $5 million in donations in just 5 days and that number eventually topped $13.1 million. Compensation will be available through the Federal Disaster Assistance Program, but home insurance is unlikely to cover costs for damages. Recovery may take many years and after the flood, the hard questions begin. Samantha Warwick (2013, F6) writes, reflecting on the Alberta floods: "Water is unpredictable. The flood is an unexpected reminder that no matter how organized and safe and cautious we are, we are never fully in control of our lives. Water is a cleansing element too—a beautiful, renewing and energizing force that also has the power to flood, storm, freeze and drown us."

Have you experienced a flood? Many people answer "Yes!" to this question, even if the flooding conditions were a brief encounter while travelling or involved only a basement flood, caused by poor drainage around their home: "When I was a child, Duncan, my father, built a river in our basement. There's always been one flowing underground, he said. He was going to tap into it, tame it, direct it. Water always seemed to be coming into our basement anyway. We had constant floods … Duncan, despite initial assertions that we were on terra firma, decided that we must accept our condition. Embrace it even" (Layton 2005, 9).

Flooding Events

Flooding is a natural event and is only a problem when it produces injury, trauma or death; damage to property or infrastructure; destruction of the aquatic ecosystem; or a significant inconvenience. Flooding is most often described as high water affecting floodplain land, often level land adjacent to a lake or river. In Canada, this is not the only definition. Some communities suffer flooding from ice flow jams and log jams, others from channel changes as a stream jumps its bank, often with rocks and debris in tow, flooding an alluvial fan area. Damage caused by high water, wave set-up, and erosion of land or slope failure are common, as witnessed by residents of areas bordering on the Great Lakes. Catastrophic losses can result from events such as hurricanes, dam failures or overtopping of dikes as well as flooding from a tsunami. The focus in this chapter is on fresh water flooding, not flooding from the sea. Although there are many areas subject to flooding, Canadians are fortunate compared to other countries where severe losses have occurred.

If asked whether you have read about a flooding event in a media source, you are likely to cite the Calgary event and may remember at least one of the following examples. In Edmonton, flooding in 2004 did millions of dollars of damage. Parts of the roof of West Edmonton Mall collapsed under the weight of rain and hail causing burst pipes on the second floor. On July 15, 2004, 200 millimetres of rain fell on Peterborough, Ontario backing up sewers and flooding basements. Over 4000 homes were flooded and 100 people had to be rescued from their cars. The estimated cost of the flood was over $92 million (Canada Public Safety 2011). This event alone may be compared with Walkerton. In Calgary, in June of 2005, the city reeled from the effects of the second greatest rainfall since 1902. David Phillips, the Environment Canada meteorologist who should be presented with a special award for his quotable quotes, downplays that unusual year: "People should not start building arks. It may look biblical, but it's not" (Lewington 2005, A6). The costs, however, were horrendous—$140 million (Globe and Mail 2006a, A10).

From May 22 to 27, 2005, heavy flooding affected the Bridgewater area of Nova Scotia causing evacuation of 80 people and closure of 17 roads and bridges in Lunenburg County. Flooding in Stephenville, Newfoundland knocked 76 homes off their foundations, forcing out 180 people and left that community without drinking water after a pipe broke (*Globe and Mail* 2005, A11). Flooding in New Brunswick in 2008 was the worst since 1973, although the damages are less than an earlier event given completed protective measures (*Globe and Mail* 2008, F2).

In May 2011, the Canadian Forces were called in to help residents along the Richelieu River in southern Québec. Before the river receded, 3000 homes were flooded and 1000 people were evacuated. The cost to the province from this record-setting event was $5.6 million (Paperny 2011, A6). In 2012 heavy rainfall brought flooding to Truro, Nova Scotia causing $3.5 million in damage costs. It is not surprising that Public Safety Canada ranks floods as the "Number One" cause of property damage (Canada Public Safety 2010). Yet these events pale beside the costs of Katrina (U.S. government costs estimated at $150 billion (Trumbull 2005)) and the tens of billions expected from Hurricane Sandy in 2012. But if you have experienced a flood, you will know that even a small event brings with it the same misery as those more publicized.

Two other examples, explored in more detail, show what happens when floods occur and, in particular, the response to floods. The first event took place in the Oak Hills area of Kamloops, BC where 801 urban-size lots were approved, although not all were built upon. The flood event began on May 28, 1972 with seepage taking place under the dikes, soon cleared by pumps. When the seepage continued, the owners of the land raised the dikes. But a few days later, a section of dikes collapsed and water inundated the development damaging some 65 mobile homes and 125 single family dwellings. Observations of this event, replete with political bias, are gleaned from the debate on the introduction of the BC *Land Commission Act* to preserve agricultural land:

> MR. G.H. ANDERSON (BC Hansard 1973, 1734-35): ... the decision has been made and "Soak" Hills is going to return to being Oak Hills, even though there are no hills there and even though they haven't yet planted the oaks.

> This was a piece of agricultural land, Mr. Speaker. It was a place where they used to winter cattle. In the spring when it flooded it made no difference because the cattle had been moved up into the hills and the ground was saturated with a growth from some hay that was cut and stored for winter. This was a piece of agricultural land that every resident in the area fought against seeing developed as a housing project - to no avail. As a result last spring, with the little heap of dirt that was piled up, the dike broke. If it had broken at 4 o'clock in the morning instead of 4 o'clock in the afternoon I'm sure, Mr. Speaker, there would have been lives lost and it would have been a much greater tragedy than it was. Over $1 million has been paid out in reparations to the victims of that flood absolutely unnecessarily because it never should have been developed. Over a million dollars is going to be paid from taxpayers' money to build a new dike to protect the people that live there. This is the kind of a price we pay for the former administration's treatment of what is agricultural land.

And later:
> And we all know about Oak Hills, where the dike built under the previous government collapsed in the first high water and flooded the residents out,

and this government stepped in and built a suitable dike that will protect those people for many, many years to come. It's up to federal and provincial standards - not simply a pile of dirt that we got under Social Credit, and some of that dirt we're still getting, mostly in this chamber. But as the Attorney-General has said: "Watch it, because he who throws dirt loses ground." But now we have a situation where we have proper lots, properly protected and put on the market by this Minister (Anderson in BC Hansard 1975, 2072).

In the wake of the Oak Hills flooding event, and with a speed reminiscent of passage of the Lower Mississippi Flood Control Act of 1928 in reaction to the great flood of 1927, the BC government, just prior to an election, passed new legislation governing subdivision on floodplain lands in the province.

An example of a much larger flooding event comes from Manitoba. In an informative article in *Canadian Geographic*, Pindera tells the story. He starts by reminding us that May 1997 was not the first flood, for native people had told of the great flood of 1775. New settlers to the area probably did not believe them until, in 1826, the 100 metre wide Red River spread across the prairies. It happened again in 1852 when "[d]welling houses and barns were floating in all directions like sloops under sail, with dogs, cats and poultry in them" (Ross 1852 cited in Pindera 1997, 55). Damage was estimated at $2 million in today's dollars. And it happened again in 1861 and again in 1948. Between these two periods, Winnipeg continued to grow and land drainage programs brought more farmers to the Red River Valley. In 1948, farmers' dikes protected low lying areas. In 1950, 80,000 people were evacuated from the city, 20,000 from rural areas and flood damages cost $606 million.

From then on, a continuous effort built structural works to protect the area: a 47 kilometre long channel (floodway) which diverts the Red River around Winnipeg, the Shellmouth Dam and the Portage Diversion built upstream to control the flow of the Assiniboine River, dikes built around Winnipeg in the form of raised roads and railways, and permanent dikes built around 8 towns as well as around 700 rural homesteads (Pindera, 57; Shrubsole et al. 2000, 12).

In the 1979 flood event, only 7,000 people had to flee their homes. In April 1997, with a near record snowpack of 250 centimetres just beginning runoff stage, a storm dropped an additional 50 to 70 centimetres of snow and freezing rain. The result: the floodway, dams and dikes reduced the crest of the flood which would have been five times higher than the 1950 flood, 28,000 persons were evacuated, downtown Winnipeg only came close to being flooded, and $150 million in flood damages were paid out. In the aftermath of this flooding, inadequacies identified included: the flood warning system, public information programs, people ignoring flood warnings, and delays in local-level response due to misunderstanding of financing for flood fighting or inexperience of responders (Shrubsole 2003, 33). Since that time the federal and Manitoba governments have introduced programs to raise building foundations and build dikes. Other measures include sealing of some groundwater wells, upgrading of GIS and topographic data, flood forecasting improvements, and geophysical research into the historical pattern of flooding (Shrubsole 2003, 36).

In 2005, Manitoba expanded the floodway around Winnipeg at a cost of $665 million. Without this intervention, what was called "the worst flood of the century" in 2009 would have seen even greater damage costs (Lambert 2009, A9). All this should have been enough for the Province of Manitoba, but in 2011, the Assiniboine River reached its highest level since 1923. Before this event was over, it was necessary to breach a dike at the aptly named Hoop and Holler Bend (fortunately causing less damage than expected) to protect other areas, and across Manitoba nearly 3,000 people were evacuated from their homes. Farmers simply gave up on planting their crops, prevented by flooding during the short time available. Again a response came in the rapid construction of a $100-million emergency channel to bring down water levels on Lake Manitoba and Lake St. Martin throughout the winter. A second waterway at a cost of $60 million is to be built (Canadian Broadcasting Corporation Manitoba 2011).

When predicted flood events are combined with a climate change scenario, the results can be frightening. By the 2050s, the land at risk of ocean flooding and storm surges across Canada is about 25 percent larger than Prince Edward Island, including 16,000 to 28,000 dwellings (National Round Table 2011, 67, 68). Rising sea levels adjacent to diked areas could affect dike maintenance, land use planning, and in un-diked areas, natural ecosystems. Sea level rise and increased precipitation will also affect homes, docks and port facilities and may raise groundwater levels. In upland areas, greater precipitation may trigger more debris torrents.

Floodplain Management: Reacting to the Flood Hazard

The previous discussion illustrates many of the elements found in flooding events: unexpected or little advance warning, human reaction (grief, search for solutions, reparations) and various actors (flood victims, other residents of the area, government officials or politicians). Noted American geographer, Gilbert F. White, observed these factors, particularly all the money poured into what he called "adjustments"—single-purpose dams, levees, and seawalls built after a flood (Kates 2011, 8). These approaches, he felt, give a false sense of security. In his doctoral dissertation, he suggests that there must be other ways to protect against flooding, not least of which is locating activities outside of the floodplain (White 1945). He stresses the importance of attention to the local or regional context and community empowerment in decisions regarding floodplain lands. White's work was expanded and refined to give many choices that modify the flood loss, modify damage susceptibility, accept the loss, or do nothing. Floodplain managers now speak of the choices between risk elimination, risk reduction, and sharing the risk (Burby 1999, 251).

Much research has explored the "hazard-response relationships"; what and who influences decisions (Penning-Rowsell 1996, 85-86). It shows how people perceive a flood hazard and how they will only respond rationally within the extent of their knowledge of the risk, and then only if they can afford to do so. Secondly, researchers evaluate tools to reduce flood damage potential, their cost, and means of implementation. Government intervention is accepted in view of public ignorance of risks, failure by individuals to assess risks, and the public good belief that it is better to protect against flood damage than to bear the losses. Sometimes there are just few alternatives to living on the floodplain. Today, there is need for a complex array of activities, integrated action (see Chapter 11), and recognition that each community is different (White 1994, 5). British Columbia's Lower Fraser Valley illustrates well the role and importance of government interventions to minimize flood damages and loss of life.[71]

Minimizing Flood Damages

The cause of flooding is similar in the lower Fraser Valley and in the Red River Basin. Winter precipitation occurs in the form of snow, possibly creating high soil moisture content. With the warmer temperatures of spring, the freshet starts in tributary streams and then the annual runoff from the 233,000 square kilometres Fraser Basin begins. The snowmelt will happen more quickly if there is a rapid rise in temperature or heavy rain. Decimation of lodge pole pines by pine beetles in the upper Fraser watershed contributes to this early start. If there is a cold spring and cool early summer, the snowmelt may be delayed and suddenly the runoff occurs over a very short period. Recent years have featured more flashy storms with greater levels of precipitation. Perhaps attributable to climate change, these often bring localized winter flooding, particularly to the lowest agricultural areas.

British Columbia's greatest flood in history took place in the lower Fraser Valley in May of 1894, before the dikes were built. Despite high waters, damage was limited as there was little development in the affected area. Another major flood occurred in 1948; damage costs to agricultural, commercial, residential and industrial development would have been close to $150 million in today's currency. Dikes failed or were overtopped. This flood caused the evacuation of 16,000 people, damage to 2,300 homes and 1,500 residents were left homeless (Fraser Basin Council 2005b). Since then, there have been a number of close calls, warning that the next big flood is "just a matter of time". Climate change exacerbates this risk as dikes were not constructed to withstand anticipated levels. Should this occur, and with major dike failures, 300,000 people could be directly affected (Fraser Basin Council 2005b). Even with this brief history, it is obvious that people will seek protection in various ways: structural protection in the form of dikes and dams, land use planning to avoid floodable areas, floodproofing of individual buildings, and disaster relief. Public education is an essential part of each of these.

Structural Protection

After the 1894 flood, farmers moved fill to the river's edge, creating a first level of protection and the base for today's dikes. After the 1948 flood, an emergency program substantially rebuilt the dikes, continuing the reliance on this type of protection (Fraser Basin Management Board 1994, 1). But there are other options besides diking: dams and reservoirs to reduce flood levels or the creation of new channels or diversions (de Loë 2000, 5). On May 24, 1968, after 8 years of study and 5 years of review, the federal and British Columbia governments instigated the Fraser River Flood Control Program with the first task to consider these options.

A study ruled out major diversions on environmental grounds, a story well told by Matthew Evenden in his book *Fish versus Power* (2004). Of particular concern is potential introduction of the parasite-carrying Arctic pike from the Arctic drainage basin and, similarly, the Pacific parasites introduction into the Arctic basin (Fraser Basin Management Board 1994, 23). The final choice was to repair, strengthen and construct dikes. The program concluded in 1995 having spent $143 million, constructed 250 kilometres of dikes, and protected 55,000 hectares of floodplain to the 1894 flood level. The province originally took responsibility for dike maintenance, but this has since been passed to local governments. Now the dikes are acting as a catalyst for a project showcasing the Fraser River. In 2009, the provincial government gave funding of $1.25M each to Metro Vancouver and the Fraser Valley Regional District for the "Experience the Fraser" project: a system of inter-regional trails along the dikes, river-based infrastructure and heritage features extending from Hope to the Strait of Georgia.

A dredging program occurs in certain areas. In addition, a debris trap, constructed in 1979 along the north shore of the Fraser River between Agassiz and Hope, catches as much as 2,400 truckloads of woody debris in a year. This debris would cause significant damage if allowed to travel downstream during the spring freshet (Fraser Basin Council 2006a). British Columbia's public safety ministry takes over management of the debris trap in 2011 and the Port of Metro Vancouver is contributing the annual $0.5M operating costs.

After the 2006 floods on the Chilliwack River, the province returned to a historic single-purpose approach and found $1.5 million to repair the dikes. An extraordinary snowpack in 2007, coupled with a 2006 study showing that the flood profile[72] is higher than originally-calculated in 1969, suggests possible widespread dike overtopping and failures and spurred new action (Fraser Basin 2006b). The provincial government promised $33 million for upgrading of dikes and the federal government contemplated a $300 million infusion of funds, the latter a component of recession-related spending.

Land Use Policy and Floodproofing

The Fraser River Flood Control Program included BC's agreement to encourage local land use planning, zoning and subdivision approvals to diminish flood losses. In general, floodplain areas are to be kept free of urban development except where committed through early settlement.[73] Developments that would increase the density or change the uses of land in any floodplain area, outside of certain designated historic settlement areas[74], are to be floodproofed.[75] In the case of the lower Fraser Valley floodplain, this means raising buildings from 3 to 12 feet depending on location.

Over time, these approvals would mean an immense administrative task. Fortunately, the task is significantly less since the *Agricultural Land Reserve Act* (1973). Under this Act, it became much more difficult to change floodplain lands in agricultural use to urban development. While land use policy and floodproofing are effective in many areas, development in the originally exempted historic settlement areas continues. Local governments requested that these areas be exempt from floodproofing conditions to preserve community character and respect local zoning, but the density of development has multiplied beyond expectations. Today the dikes constructed under the Program protect an increasing investment in development—only some structures are floodproofed through structural means or on fill to the same height as the dikes, industrial and agricultural development to a lesser elevation, and development in historic settlement areas, often not at all.

Integrated Management / Locally Responsive

As the Fraser River Flood Control Program was ending, a review recommended a more comprehensive approach for the future: a management authority for the river, better dike maintenance legislation and a "user pay" model, assessment of the earthquake risk to the dikes, floodproofing in the formerly exempt historic settlement areas upon redevelopment, diking of Indian Reserves, and an information program (Fraser Basin Management Board 1994, vii). The Board reacted with the creation of an *Integrated Flood Hazard Management Strategy* (Fraser Basin Management Board 1996), a major shift away from single purpose solutions. It calls for land use planning and management of floodplain lands by all levels of government with land use control at the local level, streamlining of approval processes, adequate tools to manage development of floodplain lands, public awareness materials (particularly maps), and recognition of environmental aspects of the floodplain. Improvements to emergency response and recovery would reduce future compensation claims and identify responsibility centres.

In 2003, some provincial responsibilities for flood protection were transferred to the Fraser Basin Council. This non-government organization has arisen from the Fraser Basin Management Board and advises all levels of government as well as the private sector and basin residents (Fraser Basin Council 2005b). The province made $1 million available for the development of tools such as mapping (Fraser Basin Council 2005c). This new approach to land use management in flood hazard areas includes authority for local government to approve floodplain bylaws and for local government and provincial approving officers to determine requirements for subdivision of flood hazard areas, rather than just the provincial government.

For example, approving officers for local government may require an engineering report regarding the flood hazard and necessary flood protection, as well as the registration of restrictive covenants. Approving Officers may also modify or discharge restrictive covenants required under the original legislation. The efficacy of these changes was subject to a 2008 review by the Fraser Basin Council. A survey confirmed belief in the importance of floodplain management. Less than one-third of the respondents indicated that the new legislation's management tools are sufficient. A large majority felt that the legislative changes had not improved management or regulatory effectiveness and efficiency and local governments have not undertaken updates to flood construction levels or flood hazard mapping since then. The report asks for an enhanced provincial role (Fraser Basin Council 2008, 2-4).

Successful initiatives of the Fraser Basin Council include mapping and public information on floodproofing options for historic settlement areas (Fraser Basin Council 2003). The latter recognizes that raising buildings can be costly and proposes new flood resistant design concepts; for example, wet floodproofing, the use of building materials that can be flooded without suffering major damage, provision of adequate storage space above the flood construction level, and the design of openings to avoid the build-up of hydrostatic pressure. The next section discusses another aspect of flood preparedness and response, where an integrated approach is key, the preparation for and activity during a flood event.

Flood Forecasting, Disaster Planning, Flood Insurance

Recent flooding in Calgary shows the incredible response to emergencies by volunteers and charitable organizations. Other examples include the Peterborough Flood Relief Concert of 2004, 3000 volunteers who offered to help clean up after the Richelieu River floods in 2011, and the leadership provided by the Red Cross in devastating events around the world. Emergency management is also a comprehensive system involving government agencies, private sector employers, and the general public. There are many activities: preparedness programs such as flood forecasting, flood warning, and disaster planning—response programs when a flooding emergency occurs and then recovery programs, as well as support for prevention outlined above (BC Emergency Management 2012).

Flood forecasting and flood warning are usually the responsibility of provincial governments. Forecasts rely on measurement and interpretation of snow packs and river flows as well as Environment Canada's weather forecasting. Disaster planning is a tremendous organizational effort at all government levels including plan preparation, training exercises, and the development of communications systems. Recovery and reconstruction programs, again at all levels of government and with the flood victims, ensure that their structures are safe for re-entry. Then comes the tedious work of cleaning and repair. Much rests on the provision of financial assistance. In the event of a large scale disaster there is a formula under which the Federal Government contributes to disaster relief (Canada Public Safety 2012).[76] Otherwise, provincial governments are responsible; these demands continue to grow. British Columbia increased its individual disaster relief limit to $300,000 when costs to replace losses rose 150 percent since 1993 (BC Public Safety and Solicitor General 2005). Such figures only enhance the importance of appropriate prevention programs and emphasize the potential for damage.

This chapter would not be complete without reference to the Canada Federal Flood Damage Program, most active from 1975 until the 1990s under the *Canada Water Act*, and an important step forward in overall floodplain management. Only the Yukon Territory and Prince Edward Island did not participate in this program that produced very high quality maps to help communities better understand the flood hazard.[77] Flood risk area designations determine floodproofing requirements and affect damage payouts where new development is not adequately protected. Neither federal nor provincial funding is made available for development on the designated lands. The program is now much weakened from its original form and this is decried: "The possibility of squandering the significant gains that have been made over the last few decades is real - especially in regions that have less commitment and capacity to treat the floodplain appropriately" (de Loë 2000, 367; Shrubsole et al. 2003).

Is flood insurance effective and should it be available? In Canada, residential insurance policies do not cover flooding events other than, in some cases, flooding from storm water sewer back-up.[78] Industrial and commercial concerns may be able to purchase insurance, particularly for temporary closure and damage. Canada has no program similar to the US National Flood Insurance Program which began in 1968, designed to share the risk of flooding. That program reallocates the cost of flood losses to persons living in floodplain areas through insurance premiums, although not without assistance from the Federal government; discourages development in floodplain areas; and maps these areas (Platt 1999, 7). Twenty thousand communities in designated floodplain areas now participate by adopting floodplain management bylaws to ensure flood damage reduction (United States Federal Emergency Management Agency 2008). In order to get financing to buy, build or improve structures in Special Flood Hazard Areas, you require flood insurance. If a structure is in place before a community joins the program, a rate subsidy is available.

According to the US National Flood Insurance Program, it is self-supporting for the average historical loss year. However, the Program has borrowing authority from the US Treasury to cover extraordinary years; this special funding must be paid back with interest (US FEMA 2004). For example, the current debt of the flood insurance scheme was $17 billion before Hurricane Sandy and has caused reforms to be passed so that homeowners pay a more realistic premium (McKenna 2012, B8). A report from the Canadian insurance sector (Insurance Bureau of Canada 1994), suggests that there is not sufficient population affected and willing to purchase insurance to adequately spread the risk. Further, the risk is not sufficiently random as floods occur at frequent intervals. Better, they suggest, to discourage development on floodplains. Yet this position may weaken when flood insurance is seen as a market opportunity (McKenna 2012, B8; Sandink et al. 2010)

Other Considerations

Two other aspects of floodplain management are of interest as they affect communities and individuals.[79] First is the ecosystem approach. Now evident in European programs, it encourages "more room for the rivers", and "land use activities from a water perspective" (Pinter 2005, 207). A local example illustrates. The Nanaimo River Floodplain project is a success story whose partners are The Land Conservancy, the Regional District of Nanaimo, Nanaimo Fish and Game Club, Trout Unlimited, Fisheries and Oceans Canada and the Habitat Conservation Trust Fund. The floodplain conserved by this project includes 56 hectares of old-growth Douglas-fir forest—one of the four most endangered ecosystems in Canada, and sensitive salmon habitat.

Unlike dike construction that confines rivers within their banks, the ecosystem approach restores river channels to flood waters, encourages biodiversity, and improves water quality. Floodplain managers work closely with those involved in habitat protection and this, in turn, creates better institutional arrangements.[80] Unfortunately it is costly to acquire land or dismantle structural works. The ecosystem approach could be complemented by an emphasis on "No Adverse Impact". Communities and individuals use proactive community planning, ensuring that no action of one property owner will adversely impact the rights of other property owners; for example, avoiding large fill areas that will deflect flows on to other properties (Association of State Floodplain Managers 2008).

A second area of interest is the study of human response to floods, as it gives guidance about how to best undertake future floodplain management. This is not a new subject, but is as relevant as ever if we are to achieve resilient communities (see Chapter 11). A 1972 article titled "Man, Floods and the Environment" suggests that it is the perception of and reaction to a flood that determines what is done about a flood hazard (Westwater 1972). More than thirty years later, research suggests that we need to recognize the "incremental" nature of gains from floodplain management (Johnson et al. c2004). Willingness to make improvements depends on context, behaviour (values, attitudes, beliefs) and environmental knowledge about an event. The authors find that policies that raise public awareness about flood threats are important "to the extent that fear provokes reaction." George Groeneveld was responsible for a task force that called for a $300-million investment in flood mitigation after Calgary's 2005 floods. Completed in 2006, the report was released in 2012. Blogger Michael Thomas (2013) quotes Groeneveld: "If you don't get this done in a one-year window, people soon forget."

Conclusion

Flood damages will increase due to extreme weather events, intensification of development on the floodplain, and ageing infrastructure more sensitive to flood damage (Shrubsole et al. 2000, 1). This means that almost everyone is potentially involved: politicians, taxpayers, the flood victims, government officials, the private sector, the media, and all who live on floodplains. After a flood, people immediately search for solutions, from single purpose to more comprehensive approaches: avoidance of floodable areas, flood control works, floodproofing of buildings, planning (discussed further in Chapter 11), disaster preparedness and public education. Responsibility for implementing these solutions now falls to local and regional governments, but funding is very much the purview of senior governments.

There is an ever-present reality that senior governments will react to crises first—

fighting floods and providing relief. Taking urgent action in a flooding situation is both essential and more politically palatable than other options. Addressing reporters on May 28, 2011, Québec Premier Jean Charest expressed his sympathy for the flood victims of the Richelieu Valley: "No one expected this to last five weeks. No one expected a flood that would bring as much damage as the one we have now," he said. "We are aware of the fact that all of this has an impact on the morale of the people" (Pope 2011). At these moments, immediate solutions are essential and altruism comes to the fore.

Perhaps such single-purpose approaches will change. In June 2011 Prime Minister Harper toured the flooding in Québec and announced preparation of a national disaster-mitigation strategy to include raising homes in flood zones, insurance programs, burying electrical cables, updating zoning and building codes, and hazard mapping. Cost sharing programs in 2011 targeted eligible structural and non-structural measures. For the future, knowing that society will only bear the costs of floodplain management if a risk is perceived or a crisis present, policy-makers will likely adopt an "incremental" approach based on available funding at the federal and provincial levels. The need to protect the ecosystem, recognize the human barriers to change, and react to climate change will enhance both the complexity and the importance of good floodplain management. Programs must fight the perception, held by many, that flooding will not happen or is eliminated by dikes. And we await the next great flood; the question is always the same. "As with so much else in nature's grasp, the question is not if it will happen, but merely when" (Glass 1988, A5).

THEME 3: WATER DEVELOPMENT AND INFRASTRUCTURE

CHAPTER 9: DAMS: CAN WE FIND A BALANCED VIEW?

... there in a little rill not more than a foot wide but as deep as wide, swollen by the melting snows, was a small water mill and at each revolution [of] the wheel its crank raised a small hammer which as often fell on a tongueless cowbell which was nailed down on a board. A loud tinkling gurgle as of water leaking out of the meadow. The little rill itself seemed delighted with the din & rushed over the miniature dam & fell on the water wheel eagerly as if delighted at & proud of this loud tinkling ... (*Henry David Thoreau in Journal 3: 1848-1851*).

I always thought dams a positive force; after all, they store water, control floods, and produce energy. Always, that is, until I studied the drowning of homes caused by dams. Always, that is, until Marq de Villiers' book *Water: Our Most Precious Resource* changed my thinking with its Chapter 8: "To Give a Dam. Dams are clean, safe and store water for use in bad years, so why have they suddenly become anathema?" (1999, 142). Why are these structures that I assumed so useful now intensely disliked? *Domicide: The Global Destruction of Home* (2001) gives part of the answer as it describes loss of home when dams are built. But the answer is more complex than simply the destruction of home and there are both positive and negative views of dam construction. This chapter, the first of two on the theme of water development and infrastructure, looks at Canada's experience with dams and seeks a balanced view.[81] I describe dam construction efforts and explore the benefits of dams as well as their costs, before introducing other considerations. Even though much has been learned from the past, the number of new dams proposed, often affecting aboriginal lands, points to the need for greater social and environmental awareness, local involvement, and transparent decision-making.

The Pervasiveness of Dams

By 2000 the United States had built 6,390 dams, China 22,000, India 4,295, and Japan 1,200.[82] Canada has been among the most active builders of large dams, although not as prolific. Construction occurred after the Second World War when progress was the watchword, water supplies appeared unlimited, and engineering expertise was ready to devise the necessary solutions. As of 2003, Canada had 933 large dams in operation or under construction (Canadian Dam Association 2012).[83] This figure includes 10 of the world's largest 40 dams as measured by gross reservoir capacity. Canada may have at least 10,000 dams (large and small). Of the large dams, 70 percent are for hydroelectric production, 7 percent for water supply, and 6 percent enable irrigation (Prowse et al. 2003, 2). Other purposes include controlling floods, making inland water bodies navigable, enhancing recreation, and replenishing groundwater supplies; some dams serve a combination of purposes.

The infrastructure that supports our lifestyle fascinates many people, particularly when these works are large in scale. Often of greatest interest are hydroelectric dams.[84] I too felt that attraction when, as a provincial government employee, I attended a celebration for an important completion point of the Revelstoke Dam. As we stood atop the dam, the man next to me said he was part of the construction team right from the beginning. So I asked what he would tell his grandchildren about the dam. He said: "I will tell my grandchildren that I built the Revelstoke Dam." I smiled wryly, for I was thinking something similar in that proud moment, although obviously recognizing my insignificant involvement compared to his. I think I just wanted to feel that I had a role in the creation of such a remarkable structure. But many dams in Canada are as imposing.

When the International Commission on Large Dams held its annual meeting in Montreal in 2003, delegates toured 14 projects in Labrador, Quebec, Ontario, Manitoba, Alberta and British Columbia (International Commission 2003). These projects were constructed as early as 1916 and the latest in 1991 and generate 3 to 5428.5 megawatts. Most are for hydroelectric purposes, but some provide water management, flood control and control of ice jams, erosion control, flow regulation, conservation and recreation. They are impressive in physical scale. In form, they are arch and buttress, gravity or embankment dams or power plants. Most have enormous reservoirs to store water, releasing it as needed and if available. Some are or include "run-of-the-river" facilities, having continuous flow and limited reservoir storage. Both contribute to the apparently simple task of taking the force of rushing water and turning it into electricity.

Hydraulic power's contribution to our everyday life is significant, supplying 62 percent of Canada's primary energy and the main source of a safe, clean and cheap (some of the lowest prices in the developed world) source of electricity (Canadian

Dam Association 2012). Most production occurs in half of Canada's provinces, with the largest contributors provincially-owned: Hydro-Québec, BC Hydro, Ontario Power Generation, Newfoundland and Labrador Hydro, and Manitoba Hydro. Recent deregulation of the industry encourages independent power producers (1-50 megawatts (MW)) who sell their power to electric utilities (see later section). As urbanization increases, so the industry will grow—by about 13 percent between 2010 and 2020 (Canada Natural Resources 2002). Indeed, "it is truly debatable whether Canada has yet passed its major period of large dam building" (Prowse et al. 2003, 1).

There are many who would agree, as news and government sources show. The *Engineering News Record* trumpets a revived plan for the $4.1 billion Thunder Bay hydropower project: "After Taking a Pounding, Hydro Is Ready for Comeback" (Wright 2003, 15). *The Globe and Mail* announces a reinvigorated project to deliver electricity to Southern Ontario (Mackie 2004, A5): three new dams will be built on the Nelson and Burntwood Rivers in Northern Manitoba. Québec announces development of greater hydroelectric capacity with an investment of $25 billion. These developments are in addition to the Eastmain 1A/Rupert Diversion/ La Sarcelle projects (Québec 2006, 9). In 2008 the James Smith Cree Nation announces that it wants to build a billion-dollar hydroelectric dam on the Saskatchewan River and sell the power to Sask Power and a diamond mine (*Daily Commercial News*). BC approves plans for the $6 billion Site C Dam on the Peace River (Hunter 2010a, S1). Newfoundland and Labrador's Premier Williams announces the $6.2 billion development of Labrador's Lower Churchill River project to be completed by 2014 in partnership with Nova Scotia (McCarthy 2010a, B1).

The Benefits of Dam Construction

"Why, in the face of such resistance, do governments still build large dams? There are two answers, one bad and one good. The bad one is that no politician can resist the prestige linked to a big dam. ... The good answer, however, is that large dams can bring substantial benefits" (Economist 2003a, 2). As the previous section shows, foremost, the construction of dams supports the production of power and hydroelectricity. This type of power creation produces fewer greenhouse gas emissions than burning fossil fuels, is renewable and is one of the cheapest forms of energy. But there are other benefits, even though some are not always as promised.

Recognizing that dams and dikes give flood protection, Ontario's Upper Thames Valley Conservation Authority today operates three major dams for this purpose. The Conservation Authority was formed in response to the flood threat; the area experienced severe floods in 1791, 1883 and 1937 as well as lesser events (Upper Thames Valley 2007). Flood control may not be the primary purpose for dams, but it is this benefit that often raises initial interest in multi-purpose projects. In 1948, Trail, BC was inundated, and in the United States, 50 lives were lost, the city of Vanport, Oregon destroyed, and $100 million of property damage done. Flood control became an urgent reason for the construction of BC's part of the Columbia River system, a system that now serves many purposes.

Many of Canada's early water management projects improved navigation and transport. The first large dam, Jones Falls, built as part of the Rideau Canal and the St. Lawrence Seaway, uses dams to control water levels. On the Trent-Severn Waterway Canal in Ontario, more than 125 dams and the lock system regulate water and maintain navigation levels; later stop log and sluiceway systems allowed lumber companies to extend the time during which they could float out logs (Friends 1996, 1). Similarly, dams constructed in Newfoundland on Pamehac Brook, with headwaters of the system diverted into the Exploits River, facilitated waterborne transport of logs to a pulp and paper mill (Scruton et al. 1998, 145).

Population growth, seasonal variations in supply, or overuse of other sources necessitates water supply storage. For example, in the Prairie provinces, there are 770 dams so virtually every major river system is affected (Canada Agriculture and Agri-Food 2006, 2). Of these, the Oldman River Dam, completed in 1991, involved intense opposition. Today the dam provides water to farmers and communities, generates electric power for 25,000 households, and creates windsurfing opportunities on the reservoir—yet still there are those who are bitter (Glenn 1999, 6-7).

The desire for economic growth is a major reason for dam development and indeed, economic growth occurs, often in areas far beyond the dam location. Damming and diversion of Long Lake in Ontario in 1939 to provide cheap power to southern metropolitan areas in Ontario and Québec, as well as Michigan and New York, is an early example (Quinn 1991, 148). Progress is frequently believed to be paramount and there is enormous prestige accruable to its servants. We need only remember W.A.C. Bennett, premier of British Columbia from 1952 to 1972 and his promotion of the Columbia River Project and Premier Jean Lesage of Québec who, under the banner "Maîtres chez nous" initiated the James Bay project.

Dam construction promises enticing regional development. Evenden (2004, 121) tells of the number of positive submissions regarding dam construction received by the Post-War Rehabilitation Council, a group of politicians who toured British Columbia in 1942. Guy Constable, a BC mining engineer and one-time Secretary of the Creston Dyking District, would be in sympathy with these positive comments. Constable involved himself in many facets of the life of the Creston Valley and Kootenay District including promotion of the Columbia River Project. The Creston flats have exceptionally rich soil, but floods endangered local farmers' prosperity. The Libby Dam would be the solution. Constable displays his trust that regional development would result from dam construction: "This country was built up by giving each place its place in the sun. I am not going to say that the whole of British Columbia will benefit, but we are entitled to our place in the sun" (Constable 11/1). His untiring efforts represent regional interests that focus on the question of which, not whether; dams should be built, often overwhelming protests that arise from the affected area's residents. Today, similar optimism continues with power proposals expecting to reap new riches.

Economic benefit in the immediate area of dam construction accrues mainly to construction companies, their employees and consultants. The Revelstoke Canyon Dam contributed $312 million (1985 dollars) in income; nearly 70 percent the result of project construction. The community received water system and community centre improvements, housing, a sewer system, road improvements and fire protection (Skaburskis 1988, 669). Businesses in the small community did benefit, but their good fortune was short-lived. In 2012, it is a similar story as construction of a fifth turbine (500 megawatts) at a cost of $230-$250 million provides 380 person-years of direct employment (BC Hydro 2012b). Otherwise the community must rely on tourism and recreation for economic development. The mayor of Fort St. John contemplates the construction of the Site C Dam on the Peace River: "Are they interested in becoming part of the community and contributing like all the rest of us … Or are they going to sit on our outskirts and when the guys get a night off, they are going to come into town … ?"(Hunter 2013, S5). Or as observed about another project: "Perhaps, NASA will one day call on Hydro Québec to organize the shuttle of workers to the space station when it is built, since the workers on the La Grande Complexe may as well be on the moon; they arrive, work and leave on shuttle flights" (Moses 1999, 6).

Finally, there may be a few environmental gains including the provision of a barrier to upstream migration of detrimental non-native species, habitat for rare flora and additional wetlands habitat (Donnelly et al. 2002, 12). The benefits of dams are many; unfortunately there are also less positive aspects.

The Problem with Dams

The problems caused by dams are well-documented: cost both to construct and to maintain; loss of home and landscape, potential forestry and agricultural production, and environmental values (disruption of ungulate habitat); shoreline erosion; ageing of dams and potential dam collapse.

Economic Considerations

Despite the economic gains from dams, there are costs. Hydroelectric dams often prove more profitable than other forms of energy generation as "fuel" is low in cost. Dams built for water supply are less profitable; an inadequate price is often paid for water, particularly from the agricultural sector (Swiss Re 2003, 7). Large dams require a significant long term capital investment. Construction and operation also involve a certain degree of risk, an important consideration from the point of view of insurers (Swiss Re 2003, 2). As discussed below, dams have a limited service life and need continuous maintenance. Together these factors lead to government participation in initiating and continuing such projects. Misinformation from megaproject proponents and lapses in transparency on the part of decision-makers are also concerns, resulting in a public ill-informed of the financial risks and an apparent lack of consideration for the public interest (Flyvbjerg 2003). As long as the light turns on or water rushes from the tap, no questions are asked. Yet there are serious social and other costs of dam construction.

Loss of Home and Sense of Place

"It is more than the land you take away from the people, whose native land you take. It is their past as well, their roots and their identity. If you take away the things that they have been used to see, you may, in a way, as well take their eyes" (Dinesen 1972, 375). Dinesen's quote echoes in my mind from an earlier discussion on this subject. In *Domicide: The Global Destruction of Home* (Porteous and Smith 2001, 12), we develop the concept of "domicide: the deliberate destruction of home by human agency in pursuit of specified goals, which causes suffering to the victims." By studying records and listening to residents of BC's Columbia Basin affected by dam construction, the effects of domicide are found to be physical, social and psychological: loss of environment, identity, community, livelihood, a beautiful place, familiar surroundings, final home (graves), land, security, initiative and health. The loss of historic continuity, the sense of being rooted, was also felt.

Anne DeGrace writes in her novel *Treading Water* about the small community of Renata that disappeared with dam construction in the Arrow Lakes area of the Columbia River Basin (2005, 252). In her words: "After Cheevers [the assessor] left, I walked through the house that was once my mother's and father's, touching things: the neat hem of the kitchen curtains, sewed by my mother's hand; the bottle my mother always kept on the kitchen windowsill that cast a watery blue glow across my hand as I reached for it. I sniffed at its rim, but there was only the scent of dust. What had it held?" DeGrace's work is the reality of "memoricide", human memory expunged. Loss of property rights is also a significant factor for those who lose both home and livelihood. Porteous and Smith (2001, 179) suggest that residents whose homes are in the way of dam construction unconsciously attempt to achieve the true monetary value of what Million (1992, 153) calls their "embodiment of identity".

The effect of dam construction on BC's Cheslatta T'en First Nation and the James Bay Cree gives another perspective. Such developments affect people who may be "still living in traditional territories [and] are particularly vulnerable to artificial changes in water distribution and quality imposed by external development" (Quinn 1991, 142). Here loss is of personal and group identity and sense of place, rather than just the construct of home: "First Nations tend to have stronger spiritual and emotional connections to home places ..." (Windsor and McVey 2005, 147).

Alcan's Kemano Project in BC grievously affected the Cheslatta T'en First Nation. The Cheslatta lived on 17 reserves, an area of about 1,052 hectares, and were displaced by flooding for the Nechako Reservoir. Now, more than fifty years after their forced departure, what is lost is more evident: a prosperous, self-sufficient life including traditional hunting, trapping and fishing; small villages, communal pasture and trails and wagon-roads; a church built by the Cheslatta of hand-hewn timbers; and their graveyard that has flooded every year (Windsor 2005, 154-56). These are physical losses and loss of a way of life. Their familiar landscape and place of memories no longer exists and they are forced to live in the ancestral lands of others. Sense of place and identity is affected and members of the Cheslatta T'en First Nation were said to suffer from alcoholism, depression and suicide (Smith 1995, 122). The experience of the Cree of James Bay echoes this account of loss and a subsequent, long-awaited rebuilding.[85]

In the 1970s, development of Québec's water resources became the key to prosperity and sovereignty. The four dams comprising the Grande River hydroelectric project were to take advantage of diversion of 11 rivers flowing into eastern James Bay. Hearing about this proposal through the media, the Cree (numbering about 6,500 people) began what was to be a lengthy fight through the courts and in negotiation with the Province of Québec to try to save their traditional lands. These lands were their home for thousands of years and supported their way of life and identity, their practice as subsistence hunters, and their culture, yet there was no legal requirement for public review of the proposed project. The review ultimately occurred in the form of lengthy court proceedings.[86] Legal and negotiation processes changed the Cree's "ability to define the cultural/environmental context that had made sense of their unique relation with James Bay for thousands of years" (Carlson 2004, 2). They also formed the equivalent of a social impact assessment, previously neglected by the project proponents (Brody, accessed 2012, 9).

While the Cree and expert witnesses enunciated the meaning of their land in terms of the search for food, areal extent, and geographic characteristics, it was much more difficult to express loss in terms of the spiritual meaning of the land. Matthew Coon-Come, then Grand Chief of the Cree, writes: "A project of this kind involved the destruction and rearrangement of a vast landscape, literally reshaping the geography of the land. This is what I want you to understand: it is not a dam. It is a terrible and vast reduction of our entire world" (Coon-Come in Carlson 2004, 2). The Cree's relationship with the land was "one of active and effective hope with an environment that was subjectively involved with individual hunters," a relationship unique to each hunter (Carlson 2004, 10). Peter Hutchins, a junior counsel for the Cree, explains this relationship: "The land is the centre of [their] existence. It is peace, it is tranquillity, it is harmony with nature. … And this is a communal concept. The land for them is a home and it is also a garden and it is a garden in two senses." He spoke of the garden as a place to gather resources for daily living and the garden as a place of cultural origin (Carlson 2004,18). A study of the social consequences of hydro development for the James Bay Cree suggests that people who lived on Fort George Island (relocated to Chisasibi due to erosion of their lands by development-related flows) subsequently saw their community suffer from alcohol and drug abuse, family violence, suicide and juvenile crime (Neizen 1993, 2).

In January 2006 the Royal Ontario Museum opened a new display of its native collections. It was not lost on observers that the artefacts are shown through the "lens of white collectors" (Milroy 2006, R5). The homes lost by the Cree and Cheslatta T'en First Nation are seen through no one's lens. Those responsible for the power projects did not see these lands as of importance. Those whose sense of place and identity drowned with the projects will see their home no more. A report to the World Commission on Dams develops this notion suggesting that special circumstances need to be recognized in the social assessment of "wilderness developments". And it further suggests that benefits clearly accrue at the scale of province and nation, but: "How does the government calculate the earnings from export of hydro power against the dispossession and disorientation of several hundred tribal people (Brody accessed 2012, 8)?" This is now the challenge.

The Cheslatta's situation may be improving based on a number of developments since 2000. A government-funded project helped young people to clean up flood debris around Cheslatta Lake, while learning traditional skills from their elders. Cheslatta Forest Products has been formed, and the Cheslatta awarded a Community Forest Licence (Windsor and McVey 2005, 157). But a territorial agreement between the Cree, the Inuit and the governments of Québec and Canada in 1975 was regarded as ineffective in allowing for equal partnership in development by local communities (Desbiens 2004, 3). In 1986, Québec announced the second stage of its hydroelectric development, the Great Whale project, to dam and divert rivers running into James Bay and Hudson Bay. The Cree's fight continued with a costly legal challenge, but when this finally resulted in an order for further environmental and social assessments and when the major New England contracts were cancelled, Premier Jacques Parizeau ended the project.

Then, for the Cree, all disagreement apparently ended. In 2002, with the development of a treaty, lawsuits against the hydroelectric schemes ceased; development of the Rupert and Eastmain rivers will go ahead increasing Québec's electricity supply by 8 percent and creating 8,000 jobs. The Cree will receive authority over economic and community development as well as $3.5 billion over the next half century, with these funds to increase if electricity prices rise or mining and forestry projects prosper—an example of the trend towards resource co-management (Economist 2002, 43).

One-quarter of the population of the Cree communities of Nemaska, Chisasibi and Waskaganish still opposed the project. The executive director of the Council of the Crees, Mr. Namagoose, recognizes the need for the project, but says: "I spent a lot of my life fighting to save rivers ... Yet at the end of the day, I end up losing my own river. It's difficult, very difficult ..." (Séguin 2006, A8). The project is proceeding at a cost of $6.5 billion, but Innu leaders in northeastern Québec have asked for concessions before the La Romaine hydroelectric project goes ahead (Canadian Broadcasting Corporation 2010a). In January 2011, the Innu of Uashat Mak Mani-Utenam agreed to drop their legal action in return for a $125 million compensation package. The Innu at Sept-Îles continue negotiations (Séguin 2011, A8).

Ageing of Dams and Risk of Dam Failure

The average age of the world's dams is around 35 years (Swiss Re 2003, 4). What this means is a need for continuous maintenance as well as reconstruction or changes to bring them up to current safety standards. Examples help to explain the costly work necessary. British Columbia's Peace Canyon Dam plans replacement of parts in the generating unit, the work valued at $322.4 million (BC Hydro 2012a). At the Conestogo Dam on the Grand River, upstream from the Kitchener-Waterloo area in Ontario, 3 of 4 gates that control flow out of the reservoir deteriorated over 40 years since dam construction (Mittelstaedt 2001, A7). Should overtopping occur, flooding and drinking water quality problems could occur. Repair at a minimum cost of $1.2 million was necessary. In 2009, more improvements are announced: upgrades to calm waters surging out of the gates so the dam's base will not be eroded (at a cost of $2.7 million) and a $20 million project to construct an emergency spillway (The Record.com 2009).

Canada is fortunate in not having experienced major dam failures resulting in loss of life, although there have been problems with smaller dams. In 2002, a gate at the Barrett Chute hydroelectric dam was opened to spill excess water. Twenty people swimming and sunbathing by the river were caught by surprise; two died (Makin 2006a, A12). Supervisory staff was cleared of negligence, but the Ontario Superior Court judge hearing the case against the employees cited "poor training, dubious procedures, and safety measures that ranged from the incomprehensible to the absurd" (Makin 2006b, A4). Common reasons for concern are extreme flood events that would cause a dam to be overtopped, static events such as seepage into the dam, erosion, ice effects or terrorism, and seismic events (Prowse 2003, 2). While the risk of dam failure is very low, the possible damage that could occur is catastrophic. For this reason, authorities set standards for dam safety, design remediation or, if necessary, retire dams.

In 1996, the Saguenay Region of Québec suffered a major flooding disaster, rated as a 1 in 10,000 year event. It caused 8 deaths, $800 million in damages, and destruction or damage to 1,718 houses and 900 cottages; in part, problems with dams were the cause. Following this event, the Nicolet Commission recommended: trained staff available during flood events, replacement of wooden floodgates with metal floodgates, and creation of an independent authority to inspect dams and ensure dam safety (*Canadian Geographic* 2007). The Québec government subsequently passed the *Dam Safety Act and Regulation*.

Extra vigilance is necessary, but there is a lack of federal regulation over dam safety; many dam-owners follow non-mandatory guidelines from the Canadian Dam Association. As of 2012, British Columbia, Alberta, Ontario (for water supply dams) and Québec have enacted dam safety regulations. The International Joint Commission, in a 2006 report (*"Unsafe Dams?" Seven Years Later—What Has Changed?*), finds an increased emphasis on public safety and emergency preparedness since September 11, 2001, but the status of regular inspections and oversight by governments has remained largely unchanged in Canada, except in BC (International Joint Commission 2006). And even this is not enough, if monitoring and subsequent enforcement does not occur. BC's Testalinden Dam failed in June 2010 causing a mudslide that inundated a highway, homes and orchards. The province subsequently assessed 1,100 similar dams (ordering immediate remedial action in four cases); other actions include hiring of 4 more dam inspection staff and sign-posting of all dams (Nieoczym 2010, S1).

Tourism and Leisure Opportunities

While dams, where used to create reservoirs or control rivers, may be seen to encourage tourism and leisure opportunities, the reality is often less attractive. When reservoirs are drawn down, marine businesses and adjacent shore and land areas suffer. Rivers may be indirectly affected by hydro developments. Manitoba Hydro's proposals for transmission lines from new dams in the Churchill area would cut across a proposed UNESCO World Heritage site and affect the Bloodvein River. This river is "considered by canoeing cognoscenti to be one of the top runs in Canada, not because of the technical challenge of its rapids, but because of its beauty" (Shoumatoff 2005, 2).

Damage to the Aquatic Ecosystem

There is a long list of environmental problems to consider in relation to the possible impact of dams on the aquatic ecosystem (American Rivers 2005, 2; Canadian Dam Association 2005, 1). Dams stop the flow of a river, storing the water in a reservoir until its controlled release. Negative impacts include: blocked or slowed flows; reduced river levels; blocked or inhibited fish passage; obstructed movement of nutrients, gravel, and woody debris; limited public access to the river; creation of barriers to navigation; and harm to the natural vegetation. Water quality problems may include changes to oxygen levels, nutrient levels and temperature in reservoirs that may affect fish and other aquatic life. Other effects are shoreline erosion increasing the turbidity of water, increased methyl mercury levels due to submerging of soil and vegetation, stratification where layers of water of different temperatures result affecting dissolved oxygen and nutrient levels, and eutrophication in reservoirs having excess nutrients causing over-production of algae.

In addition, reduced currents downstream of a dam may expose young fish to predators for longer periods of time. Landslides may be triggered by unstable slopes and, in rare cases, there could be increased earthquake activity caused by filling of a very large, deep reservoir. Raised groundwater levels may result when water seeps from a reservoir or lowered levels occur where a dam creates a barrier to groundwater. Finally, greenhouse gases could be emitted during construction and reservoir operation and changes to the climate might occur where reservoirs have a large surface area, causing warmer winters and colder summers.

A review of environmental assessments for the James Bay Development Corporation, Manitoba Hydro and, in the 1990s, the "Hudson Bay Program", suggests that these costly and impressive studies, while undoubtedly better than nothing, did not lead to change in the way the project would be operated (Bocking 1998, 1-4). Beyond this, environmental impact assessments have been lacking in two ways. The focus on environmental assessment has not been extended to recognition of social values related to the preservation of landscape. And there has been harm due to inadequate predictions, often not realized until long after project construction. The Cree and Inuit, following the James Bay hydroelectric development, have seen changes to the stable ice cover, reservoir areas devoid of wildlife, and high mercury concentration in the reservoirs (Moses 1999, 7; Quinn 1991, 145). Reservoirs have flooded forests and marshlands, releasing methyl mercury into the water and contaminating the fish. Such was the experience of the Innu after the construction of the Smallwood Reservoir in Labrador (Canada & the World Backgrounder 2005, 2).

In Alberta, the Oldman River Dam, opposed from its inception in 1975, spurred changes to environmental management at both federal and provincial levels (de Loë 1999, 219). This is not surprising given that no attention was paid to environmental impact at the first phase of studies of the dam proposal. In the end, a Federal Environmental Assessment was undertaken in 1991 on a dam that was half complete (de Loë 1999, 226, 230). Twenty years later, the Oldman River Dam Environmental Advisory Committee recommends: attention to turbidity and algae growth in the river downstream of the dam, additional flows to establish and maintain the cottonwood forest, continued mitigation for wildlife habitat, efforts to ensure the "no net loss" objective for the fisheries resource, and regeneration of the riparian habitat (including willows and sweetgrass of cultural importance to the Peigan First Nation) (Oldman River Dam Environmental Advisory Committee 2001, 9-17).

In 2012, Saik'uz and Stellate'en First Nations of Northern British Columbia launched a civil suit against Rio Tinto Alcan hoping to force the company to release more water into the Nechako River. They believe that the Kenney Dam, built in 1953, has caused a decline in salmon, trout and sturgeon stocks (Hume 2011b, S1). This action is just part of a lengthy story.[87] At one stage the provincial government and Rio Tinto promised a $50 million cold water release facility to partly rectify the problem. Seeing no action on this, in 2012 the Cheslatta Carrier First Nation is proposing a $275 million facility (Hume and Hunter 2012, S1). Taken together, these physical, social, economic and environmental concerns suggest the need for new thinking when future dams are to be constructed or existing dams expanded.

Other Considerations

There are alternatives to past practice, and learnings too. We can question the need for a dam or use measures other than dams. Dam removal can right past wrongs, but may be controversial. Small-hydro capacity could be developed, where appropriate. Future dam construction should adopt international guidelines relating to dam construction and maintenance, encourage aboriginal groups' resource co-management, and learn from those who oppose dams.

Is a dam really needed? Perhaps projections of electricity demand by Hydro authorities are inflated[88] or the possible damage is too great. In the eyes of one expert: "[the] choice to dam or not to dam a river is not a choice between environmental values or economic values. Environmental values are economic values and dam construction in certain circumstances can be grossly uneconomic (Alberta Wilderness Association 2002, 1)".[89] The Meridian Dam would have flooded a 145 square kilometres area of south-western Alberta while harnessing the flows of the South Saskatchewan River. It would cost $3.6 to $5.5 billion to return only about one-third of each dollar invested. Damage could occur to the Suffield National Wildlife Area and Prairie Coulees Ecological Reserve as well as to one of the last remaining grasslands in Canada. Finally, the river system would be overtaxed (Alberta Wilderness Association 2002, 1; Walton 2001, A3).

Where dams are for flood control, long term planning may suggest better choices. Reforestation encourages retention of runoff. Floodplain management measures such as raising buildings are an alternative in some cases. Where energy production is the issue, there are renewable energy alternatives: wind turbines, solar power, and energy from landfill gas and manure waste as well as municipal waste. These are augmented by energy conservation measures. BC Hydro's *Power Smart* develops tips and techniques for the consumer including a calculator that allows you to see how much electricity you use on any appliance in your home (BC Hydro 2006). The only problem is that despite energy conservation programs since 2003, there is no measurable decrease in energy given the public desire for new electric toys (Hunter 2010b, S1).

While it is obvious that dams might require removal if a safety risk, dam removal for environmental reasons can also make good sense, particularly if there are no real economic benefits to the operation. Removing dams can be an effective way to restore rivers, save dam owners' and taxpayers' money, revitalize riverside communities and improve public safety (American Rivers 1999). Over 78 percent of dam removals occur because of dam safety or spillway capacity issues, but removals are also initiated due to fisheries concerns (Donnelly et al. 2002, 4).

Mark Angelo, a recipient of the Order of Canada, led the Save the Theodosia Coalition, a dam removal project carried out for environmental reasons near Powell River, BC. Following dam removal, attempts to restore fish to the river (salmon were reduced by 90 percent since 1956) have been difficult due to log jams. But a partnership between the Tla'Amin (Sliammon) First Nation and a Vancouver Island conservation association hopes to rectify this (Walz 2009). There are also concerns about dam removal: loss of recreational opportunities and private land, competition from the introduction of different fish, reduced river crossing, increased sediment transport and erosion, and loss of the cultural value of dams (Stott and Smith 2001); the latter concern well-demonstrated by protests over proposed removal in 2013 of Nanaimo, BC's Colliery Park Dam.

A new emphasis on independent power producers[90] brings with it both benefits and problems. About 2,000 megawatts of installed small hydro capacity operates from over 5,500 sites in Canada, about 3 percent of Canada's hydroelectric capacity of 67,000 megawatts (Small Hydro International Gateway, accessed 2012). Small hydro provides between 2 and 50 megawatts of power while micro hydro facilities supply less than 2 megawatts.[91] Detractors express concern about the impact of access roads, transmission lines, and substations associated with these projects; the need for that much power (Hume 2009, S1); and the sheer number of projects proposed effectively privatizing public rivers. But in their favour is lesser impact on the riverine environment (Hume 2006, S3; 2007, S1). Usually run-of-the-river dams, these do not flood land and can be used to power individual developments or to feed into larger hydroelectric grids. The 16 megawatt Sechelt Creek, BC generating station received the 2005 Blue Planet Prize. This project's intake and power house are designed to be unobtrusive, the salmon run is re-established, and the project includes a partnership with the Sechelt Indian Band and fisheries authorities (International Water Power and Commission 2005). Less good news was the proposed Upper Pitt River hydroelectric project. Transmission lines were predicted to disrupt wildlife habitat, interrupt the gravel flow by weirs and the salmon spawn (Atkinson 2008, S1). After a public outcry, the BC government withdrew support for the project.

At a global scale, one promising new direction was the World Commission on Dams (WCD) which, in 2000, completed the first comprehensive and independent review of large dams and these guidelines have the potential to guide future development. Their report shows that the future for water and energy resources development lies with participatory decision-making, transparent information policies, and using what they describe as a "rights-and-risks" approach to raise the importance of social and environmental impacts of dams to equal consideration with the economic dimension. They also call for sharing benefits fairly with those most directly affected, correcting errors that are identified with existing dams, and ensuring compliance by project investors, developers and operators (WCD 2000).

The United Nations Environment Program has followed up with guidelines and in 2005, the HSBC Group announced that their financial support will no longer be available to those who fail to conform to guidelines or others whose projects would harm or change natural habitats (Dalyell 2003, 47). Yet by 2012, the WCD's work is less in favour than this chapter would argue it should be. At a recent conference of the International Commission on Large Dams, one high profile participant stated: "My opinion on WCD is that it was over since it published its final report. At that time, none of the large dam-building nations supported it and nobody used it since then to build a dam. So, I do consider it now belongs to the past" (International Commission on Large Dams 2012).

Dams have many negative environmental impacts. It is this knowledge that is now fuelling the fight by people previously displaced by dams to preserve their environment along with their historic lifestyles (Shiva 2002, 66). Holistic impact assessment, fair settlements, better mitigation and overall planning, and long term assessment of the effect of change are options to improve these situations. Natural resources co-management offers opportunities. The 2004 agreement between the Cree Nation and Hydro-Québec, facilitating participation of the Cree in hydroelectric development, and in Manitoba, between the Nisichawayasihk Cree Nation and Manitoba Hydro in the development and operation of the proposed Wuskwatim Generating Station on the Burntwood River, are examples (Fortin 2004, 36). Mitigation for dams already built can be in the form of a reconciliation pact such as the one now under consideration by the St'àt'imc people of the Lillooet River area of BC. Worth more than $200 million and including a trust fund to be spread over 99 years, it compensates for damage done to the environment and the community by the construction of the Bridge River Dam (Hunter 2011, S1).

Social impact reviews need the same attention as environmental impact assessments, particularly where resettlement is a necessity. Early and extensive assessment by persons independent of engineering or environmental assessors, including the difficult consideration of intangible costs and benefits; a detailed strategy for income restoration; careful monitoring of social outcomes after project construction; and post-project evaluation are called for (Gutman 1994, 16). Also essential is alleviation of discomfort for those displaced through clear guidelines, conflict resolution processes, public availability of project-related materials, and well trained project officers.

Finally, anti-dam opposition is a subject to be explored in its own right. Of interest are both the methods used and the manner in which environmental outcomes and the plight of those displaced can change. Anthropologist Marie Roué examines the message of environmental advocacy celebrating authentic landscapes: "The rivers of James Bay must be saved because they are natural, wild, free"; but suggests there are growing difficulties in such arguments. They present a utopian view of wilderness, while modern work in landscape ecology shows how such areas are intrinsically linked to the people who occupy and create their lives there. It will be necessary "to adopt a more holistic conception of relations between societies and nature" (Roué 2003, 624-626).

Conclusion

When we consider the benefits and costs of dam construction, there are obvious advantages, but problems will take diligence, innovation, a multi-disciplinary approach and public involvement to resolve. Is there a way to put all the factors in balance? Perhaps not, but even trying shows new directions to follow. Dam builders, with their pro-development ideology, have seen nature through their own lens (Cronon 1996, 25). With new dam construction, traditional aboriginal lands with their own blend of nature and culture, will be affected. Ardith Walkem, a lawyer and member of the Nlaka'pamux Nation urges us to think differently (2007, 315): "Indigenous peoples recognize that they have a responsibility to consider how their actions will affect other people. We must not simply make decisions on the basis of our own immediate self-interest; rather, we have an obligation to preserve the integrity and intrinsic value of the waters that gave us and our ancestors life".

When dams are proposed, there are ways to improve the outcome: asking all the right questions before a decision is made to build a dam, comprehensive planning (see Chapter 11), using new understanding of social and environmental effects, and designing better mitigation processes. Alternatives such as small hydro and dam removal seem to be successful in small measure. Also important are co-management options involving those most affected, post-construction audits, and dam safety standards. Dam builders' associations now recognize the problems, but whether this will make any difference is yet to be seen. For now, our view of "dams as anathema" is not likely to disappear.

CHAPTER 10: WATER, WATER GOING EVERYWHERE

If past media attention is indicative, "water export"[92] will appear as a subject in your daily newspaper at regular intervals for years to come. Debate by the public, interest groups and politicians will display not only the complexity of hydrology, but also how complicated water management is, given strong legal and political dimensions. Fundamental to resolution of this issue is recognition of the importance of water resource sustainability. Even if only to understand the debate, knowledge about the arguments surrounding water export is important and so the current chapter is the second under the theme of water development and infrastructure.

It is daunting to tackle this discussion when others have dealt with the subject so well (Bakenova 2008; Bakker (Ed.) 2007; Barlow and Clarke 2003; Boyd 2003; Quinn 2007; Windsor 1992, 1993; and many legal experts). Yet the subject has a continuing fascination, fed by media coverage over many years; a search of Canadian periodical indices from the 1960s to 2001 discovers over 1100 references (Bakenova 2008, 283-291). The main focus of this chapter is on water export (concern and response), but the chapter also contains sections on the role of the media and on the growing water bottling sector. First, background to the discussion, is the potential for water conflict.

Water Conflict

The critical importance of water resources is a consistent message. *Fortune* in May, 2000 is oft-quoted: "Water promises to be to the 21st century what oil was to the 20th century: the precious commodity that determines the wealth of nations" (Tully 2000, 342). The presence of water has become a geopolitical[93] issue. Both vital and often scarce, water is a strategic resource. Water's conflict geographies may be on the basis of scale: large-scale, basin to village, and gender-based; or by type: conflicts over integrity, access, or use; or involving social, political, cultural and economic aspects (Harris 2002, 744). In this chapter, the focus is mainly on conflicts at a large-scale or basin level and at a community level (bottled water). The focus is also on those based on access and aspects other than integrity or use, although the latter two may be part of the concern. Social, political, cultural and economic aspects loom large in this discussion.

The conflicts are hopefully potential, but could be very real if the use of water by one country causes water scarcity in another. The Nile Basin, the Tigris-Euphrates system, and the Jordan River and groundwater sources exemplify such situations. The conflicts could be very serious if destruction of the infrastructure and facilities that deliver water results. Predictions of future world water shortages add to the potential: more than 2.8 billion people in 48 countries will face water stress or scarcity conditions by 2025; by 2050 that will rise to 54 countries (4 billion people - about 40 percent of the projected global population of 9.4 billion) (UN Environment Program 2002, 2).

On World Water Day in 2001, the World Health Organization announced its policy on "human right to water" (World Health Organization 2001).[94] Water should be considered a public trust and protected as a fundamental human right under international human rights law. The right to water is recognized in General Comment No. 15 of the UN Committee on Economic, Social and Cultural Rights in 2002 and is now referred to in different acts, declarations and conventions. There should not be any argument about supplying water on humanitarian grounds. Canadians may know this, but water is not a human right in Canadian law and, in 2008, Canada's UN representatives voted against a resolution to include water as a human right in references of the UN Human Rights Council, thus defeating the resolution (Diebel 2008). Besides, Canadians' interest is elsewhere, focusing on diversion of water from Canada to the United States to serve a ten times greater population.

What are the Reasons for Concern?

Water export raises public concern given the apparent inefficacy of international treaties or law to resolve conflict, the growing environmental threat, the memory of past proposals, and a continued presence in media coverage. It is pleasant to believe that if problems arise, everyone will share water or, if there are serious water conflicts between Canada and the United States, the International Joint Commission (IJC) will solve them. The IJC has been described as the body created by the United States and Canada "to shift conflict about water issues away from their national agencies into an arena dominated by experts and insulated from pressures for narrow national advantage and short term political benefit" (Doig 2002, 12).

Be that as it may, the *International Boundary Water Treaty Act of 1909* gives the IJC authority over a broad mandate relating to transboundary waters. In this role, they have been called on for assistance on more than 100 occasions (Schornack and Nevin 2006, 2). The Commission has regulated water levels (Lake of the Woods), held hearings on major dam proposals (Columbia River system), and made recommendations regarding water pollution (Great Lakes), water diversions (Garrison diversion), and water export, to mention only a few of their high-profile activities. But what if the IJC could not solve a water export issue? Could international law relating to transboundary watercourses assist, at least contributing a framework for decision-making.

International Legal Theories / Conventions

International legal theories and laws confront the conundrum of watercourses crossing international boundaries and "the compartmentalization of states who claim sovereign rights over resources in their territory" (Westcoat cited in Dmitrov 2002, 678). Transboundary aquifers require equal attention. Five main theories have tried to address the challenge of water use conflicts (Agnew and Anderson 1992, 228-29; Spiegel 2005): Absolute Territorial Sovereignty, Absolute Territorial Integrity, Limited Territorial Sovereignty or Equitable Utilization Theory, Limited Territorial Integrity, and Community of Interests Theory or Integrated Drainage Basin Development. The first of these, the so-called Harmon doctrine, suggests that any state may use any watercourse within its borders as necessary, without regard to downstream riparians. The others are more like riparian law, suggesting that a state may make use of international water to the extent that it does not interfere with reasonable use by other riparians. By the Theory of Community Interests, states should join with each other to make use of international river basins in a manner that will ensure the greatest benefit for all.

Translating theory into international law, the International Law Association's[95] August 1966 *Helsinki Rules on Uses of the Waters of International Rivers*, like the Theory of Community Interests, proposes that international waters be shared equitably and reasonably. In 1970, the United Nations General Assembly gave the International Law Commission[96] the task of codifying and developing the law regarding "non-navigational use of international watercourses." Twenty-seven years later, the UN General Assembly approved the Commission's rules by Resolution 51/229. The *Convention on the Law of the Non-Navigational Uses of International Watercourses* provides for the equitable and reasonable use of international watercourses and stresses the importance of avoiding "significant harm" to other states' uses. There is an obligation of riparian states to cooperate through various means such as commissions and data exchange, to protect the environment, and to use dispute resolution mechanisms (Green Cross International 2000, 44).

Despite all this, the Helsinki Rules are widely accepted but are not binding (Kaya 1998, 4). Neither the International Law Association nor the International Law Commission has lawmaking power. The UN Watercourse Convention is binding upon ratifying countries and, as they represent customary international law, on other states. While the majority of the General Assembly adopted the Convention, it has only 29 Parties as of 2012. Canada is not among these. Conflict resolution through international water law may not be the easiest course or first choice. Wouters et al. (2000, 5) suggest this may be because it is of little relevance to the world's water problems, is too controversial to adopt as part of the solution, or more likely, it "is not clearly understood". Besides searching for solutions through international agreements, agencies, and law, it is essential to constantly seek equitable and reasonable use: "The need now is to combine ecology with equity, and sustainability with justice" (Shiva 2002, 79).

Environmental Concerns

Many factors create uncertainty about Canada's water supply. As discussed in Chapter 2, while Canada has 6.4 percent of the world's renewable fresh water, this plenitude overstates reality. Sixty percent of Canada's fresh water drains to the north, while 85 percent of the population lives along the southern border and has a high per capita consumption of water, putting stress on the limited supply. In addition, pollution depletes useable sources. The environmental effect of permanent, large-scale withdrawals of water can only exacerbate these problems: "Your editorial ["Weirdness about Water"] is a prime example of the idiocy that results when one wades into scientific argument without understanding the science involved. ... As for your inane suggestion that the tap could be turned off if there were a problem, most of the ecological crises facing us today are due to the fact we had no idea there was a problem until it was too late to solve it" (Editorial 1999b; Haladay 1999, A17).

Known negative effects include reduction in water flows to riverine ecosystems, introduction of invasive species as a result of diversion projects, and taking of water that normally refreshes coastal ecosystems. Commitment to large projects may close out future opportunities. Climate change and drought are increasingly evident in decreased water supplies and changes to water quality. There is already a decline in the water levels in the Great Lakes and the St. Lawrence River. The St. Lawrence Seaway system dropped to its lowest point in thirty years in 1999 (McCarthy 1999). Lack of knowledge about future surface water and groundwater quantity and quality suggest the folly of making promises to send large quantities of water elsewhere. Yet the past has seen many proposals.

The Memory of Past Proposals and Developments

Today major water export proposals may not be frequent, but it is the memory of former schemes that causes alarm.[97] While exports may involve diverting water from shared boundary waters, there are only a few examples. The Winnipeg Aqueduct was built between 1913 and 1919 to supply water from the Shoal Lake part of Lake of the Woods to Winnipeg. During aquifer construction, First Nations' residents were forced to leave their village and abandon a peninsula they had subsequently settled (Mackrael 2013, A4). One of the largest withdrawals from the Great Lakes is the Lake Michigan Diversion. Completed in 1900, the diversion withdraws water for use in the Chicago area and then discharges it via a shipping canal into the Illinois River (part of the Mississippi River watershed), rather than returning it to the Great Lakes Basin. This withdrawal reached almost 300 cubic metres/second in 1928, but has been capped at an average of 90 cubic meters/second since 1967. Population growth, drought and climate change renew pressures for greater withdrawals (Hurley and Nikiforuk 2005, A13).

Proposals raising the greatest concern are those that would carry very large amounts of water by open-channel diversions from Canada to south of the border. Particularly famous is the $80 billion (U.S.)[98] 1964 proposal for the North American Water and Power Alliance (NAWAPA). This proposal would flood 800 kilometres of the Rocky Mountain trench and involve more than 2 dozen dams in Canada. Today, the model for the NAWAPA proposal sits in a glass-enclosed box in an office hallway of Parsons', one of the largest engineering and construction companies in the United States. "It is a 1960 state-of-the-art display with mother-of-pearl inlays, a faded colour photo and glowing rivers of light, stretching from British Columbia across Canada east to the Great Lakes, and south to the Mexican border" (Peterson 2003, 1). Is preservation in a glass-box a guarantee that NAWAPA will never re-emerge?

There are many variations on this theme. Thomas Kierans' GRAND (Great Recycling and Northern Development) Canal proposes a dike across James Bay and creation of a new fresh water lake from the rivers that now empty into the bay. The new lake's water would then be pumped into the Great Lakes Basin at a rate of 11,000 cubic metres per second. No small undertaking this, the $100 billion project expected to control fluctuations in the Great Lakes water levels; diminish water shortages in the basin, the Prairies, and the U.S. Midwest and Southwest; and end the need for any other major diversions (Gamble 1998, 1). Both damming of rivers and widespread flooding would be avoided (MacGregor 2009, A2).[99]

Another is the Central North America Water Project to transfer water from northern Canadian basins along existing prairie drainages to Lakes Manitoba and Winnipeg and then south. The Magnum Diversion proposes taking water from the Peace River Basin via prairie drainages to the U.S. Great Plains region and then south to the Missouri River. The 1968 Western Water Augmentation Scheme intends to move water from as far north as the Liard Basin through the Rocky Mountain Trench to the south. The 1966 Kuiper Diversion Scheme would divert water from the MacKenzie River Basin across the Prairie Provinces to the Souris River and then south. Finally, the Ogallala Aquifer Replenishment scheme, authorized by the US Congress in 1982, wanted Great Lakes' water to artificially replenish the aquifer, using injection wells.

The reverse scenario, bringing United States' originating water north of the border, is equally controversial. The Garrison Diversion Project, designed for irrigation and water distribution purposes[100], was first proposed by the US Bureau of Reclamation in 1957 to transfer Missouri River water into the Hudson Bay drainage system. While parts of the system were built beginning in 1968, the International Joint Commission ruled against the project in 1977. A compromise reached in 1986 under the *Garrison Reformulation Act* includes a specific requirement for consultation with Canada. But in 2000, this provision was removed and potential uses for the Missouri River water in the Red River Valley were identified under the *Dakota Water Resources Act*. In 2006, North Dakota officials revived the Garrison Diversion proposal assuring opponents that this new scheme will have no environmental impacts and will carry water in pipelines. Its benefit as a water supply for many communities is cited. The estimated construction cost is $500 to $660 million with an annual operations and maintenance cost of $1.2 million. Public meetings held in 2006 received submissions in substantial opposition including those sent by the Council of Canadians, the Province of Manitoba, and Manitoba Wildlands (Canadian Broadcasting Corporation 2006b; US Department of the Interior 2006).

Another controversial project is the draining of Devils Lake in North Dakota into the Sheyenne River, a tributary to the Red River. Flooding in this area has destroyed local homes, businesses and farmland and cost the North Dakota and U.S. governments $450 million in flood mitigation (US Geological Survey 2012). It is of interest here as a water export project related to flooding and also as, after much discussion, this project has proceeded with a number of mitigative features. Some believe that this brings into question the adequacy of the International Joint Commission and commitment of both countries to the *Boundary Waters Treaty* (Norman and Bakker 2005, 12).

Other possible means of export include greater consumption or pollution of shared boundary waters; pipelines across international boundaries, usually for local water supply; container shipments of water by tanker, railroad and truck; the export of goods that incorporate water such as beverages; or the export of goods that use water extensively in their production processes; for example, aluminium or paper (Windsor 1993, 232).

There are more esoteric means of water export such as water baggies and iceberg harvesting. The company MHWaters believes that water bags with the capacity of 0.5 gigalitres can be developed (Wood 2002, 1). Towing icebergs from Canada to the Middle East may be impractical—taking 128 days by which time the iceberg would have melted (Curtis 2002, 76). But Ed Kean, Captain of the Mottak, has a license to harvest 72,000 tons of icebergs a year off the coast of Newfoundland to be used for bottled water and vodka. He says: "... Icebergs are graded from a bergy bit to a castle. A bergy bit is the size of a Volkswagen, and a castle is a million tons or more, like the one that sank the Titanic. ... You've got to hunt for the icebergs and then wait for nature to break them up. ... or we put the stereo on full blast and play rock music. We also use a rifle." (Hajim 2005, 64). Other harvesters chop off bits of ice using chainsaws lubricated with vegetable oil or a crane and grapple (Curtis, 72). All these water export proposals suggest an issue to be taken seriously. The memory of past proposals and even some current projects keep it alive. The media play an important role in bringing the continuing controversy to our attention.

Media Coverage

While the media provide information, they also create new interest in and shape the perception of issues and events. Sources for the media include politicians, policy makers, interest groups (for example, the Canadian Environmental Law Association and the Council of Canadians) and the public. More and more, the media look to the blogosphere, Facebook, Twitter and YouTube, augmenting and sometimes changing the course of media coverage. Studying media presentations allows us to consider whether they set the agenda on a particular issue, or just simply keep it in the public eye.

The media's treatment of water export may influence public attitudes and political action. Studying reports on this topic in *The Globe and Mail* between 1991 and 2001 shows how this source framed[101] a public policy debate (Thomas 2004). In 101 articles, there were 76 news articles, 3 features, 9 columns, 8 editorials, and 6 comments. From these, Thomas studies 125 instances of framed aspects including the fascinating art of the copy editor who writes the headlines designed to grab our attention (Table 10-1). For the years when most of the debate occurred (1998-2001), there was an overwhelming (84 percent) appearance of negative framing of water export (Table 10-2). The negative frames reflect concern about water depletion,

Table 10-1: Water Export: The Headline Writer's Art
Source: (Thomas 2004)

Headline	Publication Date
New TSE listing a first for fresh water exporters	02/15/1991
BC firm may slake Californian thirst	03/01/1991
Water call goes out as debate swirls	03/01/1991
Water, water everywhere and people keen to export	03/23/1991
California cooler to BC water, Export talks put on hold	03/27/1991
Vigilant to the last drop	04/08/1993
NAFTA plus nada	12/01/1993
Permit to export Lakes water draws US fire	05/01/1998
Water exports: Canada is getting in over its head	05/19/1998
Will foreigners drink Canada dry?	05/23/1998
Canada awash in water controversy	12/17/1998
Water as commerce, water as life	08/19/1999
Ottawa's leaky water policy	10/18/1999
BC fighting leaks in bulk water plan	10/27/1999
Bulk water exports could would wash jobs away	02/12/2000
Go with the flow	03/13/2000
Our water sovereignty is at risk	12/19/2000
Canadians fiddle while California burns	02/03/2001
Don't go with the flow	03/29/2001
Chrétien lands in hot water	05/15/2001
Water exporters are all wet	05/29/2001
Flag-waving hindering a serious debate on water exports	06/01/2001
Water export policy already all washed up	06/02/2001
Exporting our water? In your dreams	06/19/2001
Porous water policy is no protection from US threat	06/21/2001

Table 10-2: Media Frame the Water Export Debate
Source: (Thomas 2004)

Negative Frames	% of frames
Open the floodgates (to future depletion of resource)	17
Moving too slowly on prohibition	13
The tap can't be turned off	12
Jurisdictional non-cooperation	12
Growing threat	7
No special treatment (to water as a resource)	6
Damned if you do and damned if you don't	4
Threat to sovereignty	3
Jobs will come to us	3
Take precautionary approach	3
Government position not straight-forward	2
Environmental harm	2
Positive Frames	
Save the thirsty world	6
Non-issue: no market	5
NAFTA is not a threat	3
Economic opportunity	2

worry that the federal government is moving too slowly on the issue, or question the result of allowing water export. While the newspaper played a significant role in the debate, whether their agenda-setting directed or shaped public opinion and caused government action, needs further study. That the newspaper portrayed the issue in a negative manner is nevertheless important as it maintains public concern and directs attention to certain aspects of the issue. Fear that water export would open the floodgates and calls for immediate government action dominated public concerns. Government control, then, becomes a major focus of interest.

Provincial Government Involvement in Water Protection

British Columbia was at the eye of the storm over water export before most other provinces. The paragraphs that follow trace the story from 1984 to 1996, attempt to outline the signs that policy changes were badly needed, and examine the difficulties faced by the government in trying to make changes. To begin, in 1984 the government adopted a policy generally in opposition to water export from interior streams, but permissive of bulk export from coastal streams (streams which emptied into the ocean) based on individual merit. In March, 1991 the government placed a water reservation or moratorium on consideration of licences for bulk removal from coastal streams. This moratorium was extended for another two years after a change in government in May of 1992. As of 1993, a patchwork of potential situations relating to water export (either in place or proposed) existed and public scrutiny of the issue grew.[102]

What tipped the balance and turned all this activity into crisis, or at least an event that provoked action, was an initiative of Multi-National Water and Power, Inc. As explained by Hawthorn (2004), the right-hand man to former BC Premier W.A.C. Bennett proposed damming the North Thompson River, a tributary of the Fraser River, diverting water to Canoe Reach in the Columbia River watershed and then allowing the water to flow south to the John Day Dam on the Oregon-Washington border. From there a pipeline would be built to take the water to California. Said the proponent Mr. Clancy: "It is not water that we would take from anywhere. It's water that was going to the sea" (Hawthorn 2004, R13).

At the time, the BC government demonstrated interest in the issue in three ways. It was concerned about the implications of the North American Free Trade Agreement (NAFTA) to the province's resources. The Provincial Round Table on the Environment and the Economy recommended against water export in 1991. A review, the *Carter Report on Water Export* (Educom International 1992), was completed. In addition, a conference sponsored by the Canadian Water Resources Association on water export allowed considerable discussion of the question (Windsor 1992). Attending this conference, I was moved by the words of Chief Kathy Francis of the Klahoose Nation in her talk "First They Came and Took Our Trees, Now They Want Our Water Too": "A creek ... not only physically and spiritually cleanses people, but it also cleanses the earth and, eventually, the sea to which it inevitably flows, if left alone" (Francis 1992, 94). I was not surprised by the title of the talk by Francis Dale, President of Citizens for Water and Power for North America: "Water, Water Everywhere: How to Get It from Here to There."

It was clear where public sentiment lay on this issue, but the dilemma was how the government should approach it. Should the government ban water export outright? Choosing this action could instigate trade-related challenges, some suggest even in relation to other natural resources. This explanation simplifies a vastly complicated situation. Clarification comes from the work of Barlow and Clarke (2003) as they describe the trade agreements and Johansen (2001) as he discusses the history of the issue from the federal perspective, as well as from others acknowledged below. In all cases, it is necessary to be aware that these are interpretations on the part of the commentators. The General Agreement on Tariffs and Trade (GATT), the 1995 World Trade Organization agreement, was created to encourage the free flow of capital goods and services across national borders and lower tariff barriers. With NAFTA, these agreements are intended as tools of trade liberalization, not as provisions for environmental protection or resource conservation (Campbell and Nizami 2001, 3).

The first concern is whether water falls under these treaties. While water may not be considered to be a "good" in its natural state, the issue is, when does it change? The answer usually is: when it is a "good" or "product". As discussed in more detail in a later section, a 1993 non-binding statement of NAFTA parties stated: "[w]ater in its natural state in lakes, rivers, reservoirs, aquifers, water basins and the like is not a good or product." Some suggest that the GATT's Harmonized Commodity Description and Coding system defines water as a good as it includes the clause "22.01 waters, including natural or artificial waters and aerated waters, not containing sugar or other sweetening matter nor flavouring; ice and snow." Others suggest that this is not a definition, but merely an organizational coding structure (Terry and Unikowsky 2003, 5-6). The federal government's *The NAFTA Manual* indicates that "[l]ike the Canada-U.S. Free Trade Agreement (FTA), the NAFTA does not apply to large-scale exports of water. As in the FTA, only water packaged as a beverage or in tanks is covered in the NAFTA" (Johansen 2001, 5). These questions of definition continue to be subject to debate.

The General Agreement on Tariffs and Trade (GATT) is also based on certain principles that cause concern including the need to treat WTO members' "like" products equally under the "most favoured nation" clause, requiring any privilege extended to another country to be extended to a member state (Article I); treating WTO members' "like" products the same as domestic products under the "national treatment" clause (Article III); and prohibiting export and import restrictions (Article XI). While the "national treatment" clause may be satisfied by ensuring the same treatment for Canadian, American and Mexican companies, quantitative restrictions on imports or exports may be eliminated, even if for environmental reasons. To respond to this, an exception permits countries to adopt laws relating to "the conservation of exhaustible natural resources if such measures are made effective in conjunction with restrictions on domestic production and consumption" (Article XX(g)). In other words, environment-specific exemptions must be applied in a non-discriminatory manner. The treaty requires "like" products to be treated the same for the purposes of trade and so water in the form of a good cannot be banned from export or import, even if produced in environmentally suspect conditions. In addition, water services fall under these rules if there is private or community sector involvement in delivery or money exchange.

The North American Free Trade Agreement (NAFTA) includes similar provisions to GATT, but Article 315 suggests that if export of a good is permitted, then it cannot be restricted in the future, unless we take similar measures in our own market. Hence the frequent admonition is that "once the tap is turned on, it cannot be turned off." Chapter 11 of NAFTA extends these provisions relating to the benefits of trade liberalization to investors. A company that has an investment in another state can bring a claim if it feels the state has infringed on its "rights". Finally, Bilateral Investment Treaties establish investor rights of corporations from each country to operate unconditionally in another country and to access each other's markets and resources. Together, these trade law sanctions and legal considerations do not make an outright ban on water export an attractive or easy undertaking.

Should protecting provincial sovereignty over the water resource be the main goal? While each province clearly has authority over the management of water, the Federal government has jurisdiction over international trade. Each province needs to stay within provincial objectives relating to management and conservation of the resource and respect First Nations entitlements.

Should the government simply protect the water resource? This goal is achievable given provincial authority to determine the degree to which the water resource is to be exploited, but the distinction must not be based on destination (i.e. domestic versus export).[103] Water rights belong to the provinces as does the right to make laws relating to water development for domestic and industrial purposes. For the BC government, resolution of concerns about bulk water export came in the form of the *Water Protection Act* ([RSBC 1996] CHAPTER 484), providing the following protection measures:

1. It is confirmed that all water, including groundwater, was vested in the Province of British Columbia.

2. No further licensing of bulk water removal from British Columbia is allowed.

3. Water removal from British Columbia is only to occur in containers of 20 litres capacity or less, unless specifically exempted. Certain licensed registrants and those registered as unlicensed registrants who were already removing water were permitted to continue their bulk water removal within clearly defined limits.

4. Large scale projects transferring water between major watersheds (as defined by the Act) are prohibited.

5. The scheme relies on the provisions of the Environmental Assessment Act to ensure that any significant water project (greater than .7 cubic metres per second) is subject to environmental assessment.

6. Amendments to the Water Act ensure that all licences are to be attached to a property appropriate to the use of the water in British Columbia.

7. While licences had been issued for "bulk shipment by marine transport", this purpose was deleted as it is a means of conveyance rather than a purpose.

8. Significant penalties for non-compliance with the act were imposed (up to $200,000 and if continuing up to $200,000 per day thereafter).

The legislation took 5 years to prepare. There are a number of reasons for this. The government was in the process of reviewing all of its water-related activities. The discussion paper *Stewardship of the Water of British Columbia* (1993), which raised the issue of water export, was considered a government-wide initiative, hence the need for significant consultation within and outside of government. The protection of groundwater was believed to be of equal importance and so required time in the preparation of policies and legislation. Finally, it is essential to "get it right" when legislation is prepared, in order to avoid future challenge. In the result, the *Water Protection Act* gives protection to British Columbia's water resources, whether the threat is from within or outside of the province.

In Ontario, another high profile case spurred action. The Nova Group received a "water taking" permit from the Ontario Ministry of Environment that would permit withdrawal of up to 600 million litres of water a year from Lake Superior for a period of 5 years. Hearing public concern, the Ontario government subsequently cancelled the permit and adopted the *Water Taking and Transfer Regulation* (Ontario Regulation 285/99) providing for the conservation, protection and wise use and management of Ontario's waters by prohibiting the bulk transfer of water out of specified water basins, among other protections; small scale export may occur with approval of the Minister. In Newfoundland, a permit was given to the McCurdy Group of Companies that wanted to export 50 to 100 billion litres of water annually from 29 square kilometre Gisborne Lake. The water would be piped from the lake to a coastal fishing outport (Canadian Broadcasting Corporation 2004). That permit was rescinded and Newfoundland adopted an *Act to Provide for the Conservation, Protection, Wise Use and Management of the Water Resources* (1999).

Other provinces adopted water protection provisions, but with differences from the BC provisions. For example, exemption to the legislation may be left to the discretion of a Minister or Cabinet. The Federal government has informed the territorial water licensing boards that licenses for bulk water removal out of major river basins will not be approved. There are also policy statements, supported by each territory (Johansen 2007). There is "thematic if not substantive convergence" between the provincial and Federal legislation and policies (Hill et al. 2008, 326). Despite these controls, many people feel that the Federal government should exert comprehensive authority.

The Federal Response

> "Prime Minister, without being too dramatic, the water export issue will be one of the most important and controversial questions of the coming decade" (Waddell 2002, 132).

What was the federal government's role in this controversial issue? While water export may have originally seemed like science fiction in the 1960s and 1970s (Heinmiller 2003, 500), possible threats still raise concern. The federal government response has taken time and, in the end, stays strictly within their legal purview. The following account begins in the late 1980s.[104] In 1987 the federal government released the report *Inquiry on Federal Water Policy* (Chapter Twelve: Water Export) (Canada Environment 1985). Despite public opposition at the Inquiry, the report adopts a cautious approach, not rejecting all exports out of hand, instead proposing legislation that could prohibit water export or permit it through a licensing system (Quinn 2007, 8). Ultimately, the 1987 Federal Water Policy declared the government's intent to assess interbasin transfers within Canada in cooperation with the provinces and territories and to prohibit, within federal authority limits, water export by interbasin diversions (Canada Environment 1987).

On August 25, 1988, the Minister of Environment Tom McMillan introduced Bill C-156, the *Canada Water Preservation Act*, partly in response to Liberal charges that the FTA poses a threat to Canada's water security. The Act's purpose was to ban water exports exceeding more than 1 cubic metre per second and create a licensing system for exports at or below the standard. That bill died on the Order Paper when the parliamentary session ended. Echoing the provincial governments' experience, the signing of NAFTA tended to constrain the scope for institutional action (Heinmiller 2003, 502). In 1993, in response to concerns that under NAFTA, Mexico and the U.S. could demand the right to Canada's water, the Office of Prime Minister Jean Chrétien produced a press release. It announced NAFTA improvements that while not legally binding include a side statement about water:

> The governments of Canada, Mexico and the United States, in order to correct false interpretations, have agreed to state the following jointly and publicly as Parties to the North American Free Trade Agreement (NAFTA):

Chapter 10: Water, Water Going Everywhere

- The NAFTA creates no rights to the natural water resources of any Party to the Agreement.

- Unless water, in any form, has entered into commerce and become a good or product, it is not covered by the provisions of any trade agreement, including the NAFTA.

- And nothing in the NAFTA would oblige any NAFTA Party to either exploit its water for commercial use, or to begin exporting water in any form.

- Water in its natural state in lakes, rivers, reservoirs, aquifers, water basins and the like is not a good or product, is not traded, and therefore is not and has never been subject to the terms of any trade agreement (Johansen 2001, 7).

In May, 1998, Foreign Affairs Minister Lloyd Axworthy promised to take measures to protect Canadian water. This came after a public outcry when it was discovered that companies such as the Nova Group were on the brink of exporting bulk water.

On February 9, 1999 federal NDP member Bill Blaikie introduced a motion demanding an immediate moratorium on bulk exports of fresh water claiming as he did that "[w]ater is something that all Canadians have a common image of in our country. Water is as Canadian as hockey, as the RCMP, as the beaver" (Blaikie 1999). His motion for an immediate moratorium and introduction of legislation to prohibit the export of bulk fresh water shipments and inter-basin transfers won the unanimous approval of the House (Scoffield and LeBlanc 1999, A6). This was not to be the first time or the last that members would introduce such motions or bills.[105] On February 11, 1999 the Federal government unveiled a three-pronged strategy relating to water export: an accord with the provinces and territories relying on their powers to prevent bulk water export from basins solely within their authorities, a study of the effect of withdrawing large amounts of water from shared boundary waters, and legislation affecting trans-boundary waters.

Until a national accord could be signed, Ottawa issued an invitation to the provinces to join in a temporary and voluntary moratorium on withdrawal and transfer of water out of five major water basins: the Atlantic, the Arctic, the Pacific, Hudson Bay, and the Gulf of Mexico. The accord was initially rejected by British Columbia, Alberta, Saskatchewan, Manitoba and Québec, with British Columbia saying that "Ottawa's wording was too lax and would compromise B.C.'s own protection" (Editorial 1999a, A16). Finally, at the Canadian Council of Ministers of the Environment meeting in Kananaskis, Alberta on November 29-30, 1999, all the other provinces and territories endorsed the accord and the jurisdictions above reserved their position pending further consideration (Canadian Council Ministers of the Environment 1999). The *Accord for the Prohibition of Bulk Removal from Drainage Basins* is non-binding, but is a statement of intent.

Secondly, with the United States, a Joint Reference was made to the IJC, to determine what happens when large amounts of water are consumed, removed or diverted from shared boundary and transboundary surface waters and shared aquifers and to make recommendations on the management of these water resources. The *Final Report on Protection of the Waters of the Great Lakes*, released in March 2000, provides a comprehensive recommendation relating to bulk water export from the Great Lakes basin (p.47): "[Governments] should not permit any proposal for removal of water from the Great Lakes Basin to proceed unless the proponent can demonstrate that the removal would not endanger the integrity of the ecosystem of the Great Lakes Basin and that:

a) there are no practical alternatives for obtaining the water;

b) full consideration has been given to the potential cumulative impacts of the proposed removal, taking into account the possibility of similar proposals in the foreseeable future,

c) effective conservation practices are implemented in the place to which the water would be sent,

d) sound planning practices are applied with respect to the proposed removal, and there is no net loss to the area from which the water is taken and, in any event, there is no greater than a 5 percent loss (the average loss of all consumptive uses within the Great Lakes Basin); and the water is returned in a condition that, using the best available technology, protects the quality of and prevents the introduction of alien invasive species into the waters of the Great Lakes."

The third part of the strategy, Bill C-6, amendments to the *International Boundary Waters Treaty Act* and to the *International Boundary Waters Regulations*, was introduced on February 5, 2001. In what may have come as something of a surprise at that time, in July 2001, U.S. President George Bush told *The Globe and Mail* that he wanted to talk to the Prime Minister of Canada about the possibility of using empty natural gas pipelines to ship Canadian fresh water to the Unites States southwest (MacKinnon 2001, A5). This prompted *The Globe and Mail* cartoonist Gable to produce yet another delightful cartoon in which Bush and Chrétien are in a small rowboat in the Canadian wilderness (c. March 28, 2001). Bush is asking how much the water would cost without the moose. The U.S. intervention failed and the new law came into force in Canada in December 2002.

The *International Boundary Waters Treaty Act* prohibits bulk removal of water from the Canadian portions of boundary waters and establishes a system of water export permits for projects that involve the use, obstruction or diversion of boundary waters.

These permits are at the discretion of the Minister of Foreign Affairs through the IJC and enhance the role of the Commission. In the United States, the *Water Resources Development Act* of 1986 prohibits the diversion of Great Lakes water unless approved by all eight adjoining state governors. Amendments to this act made in 2000 further encourage the Great Lakes states, in consultation with the provinces of Ontario and Québec, to develop and implement a decision-making standard for withdrawal and use of water from the Great Lakes Basin (IJC 2004, 4).

A step towards this end is the 2004 pact and 2005 agreement (ratified by the US House of Representatives in September 2008, not requiring similar approval in Canada) between the states and provinces that border the Great Lakes to limit export of water from the Lakes (Council of Great Lakes Governors 2001).[106] The agreement bans new diversions of water, but with exceptions for public water supply required for communities near the Basin: for example, the existing Chicago Diversion and water supply for counties that straddle the basin. In the short term though, removals via pipeline will likely be small and relate to the need of communities having shortages of water or poor quality water (IJC 2000, 17).

The signees also agree to a consistent review standard for proposed uses of Great Lakes water, better technical data collection, and development of regional goals and objectives for water conservation and efficiency. These are important parts of the agreement: "The citizens of the region now have some big choices to make. They can satisfy the needs of their neighbors by sharing Great Lakes waters according to a host of standards governed by a regional agency (Annex 2001) captive to spiraling growth outside the Basin. Or they can change their habits and export their conservation practices (as opposed to water) to these same communities" (Nikiforuk 2004, 25).

In May 2010, the federal government introduced and gave first reading to Bill C-26, an amendment to the International Boundary Waters Treaty Act and the International River Improvements Act. This bill seeks to bring all waters that flow across the international border and those that run along the border under a more comprehensive prohibition against bulk water removals. This augments existing legislation that affects waters, such as the Great Lakes, that straddle the border and introduces new powers of inspection and enforcement and stiffer penalties for violations. As with previous legislation, this bill respects provincial jurisdiction over other than boundary waters. This bill died on the Order Paper at the election, but has been re-introduced with suggested amendments as Private Member's Bill C-383 and received Royal Assent and is now law.

Interest, fed by critical comment and general public and press attention, means that political and bureaucratic activity on the issue of bulk water export will continue. In a column calling on Prime Minister Stephen Harper to keep the taps closed, the

Globe and Mail's Eric Reguly (2006, B2) notes that the Prime Minister's office received 120,000 items on bulk water export and water privatization in 2005. This was likely revived in 2011 with former Prime Minister Jean Chrétien's suggestion that there is nothing wrong with reopening the debate about sharing our water with those less fortunate (D'Alieso 2011, A3). Even fictional sources and television join in. Varda Burstyn in her eco-thriller *Water Inc.* (2004) tells how a private consortium plans to build a pipeline to ship water from Québec to the United States. Ian Waddell, a former BC Minister of Environment, penned the mystery *A Thirst to Die For* (2002) that became the CBC political thriller *H₂O*.

Other Considerations

The federal-provincial dance led to the eventual response on water export (Heinmiller 2003). The federal government first attempted to ensure the adoption of standards, Canada-wide; then, following the adoption of free trade provisions, created a "harmonization through policy interface standardization" as seen in the 1999 *Water Accord*; and finally, almost by default, entered into a process of "policy emulation" when most of the provinces adopted legislation (Heinmiller 2003). Beyond chiding the federal government for such an approach, other criticisms need attention. Not least of these is the Council of Canadians' unrelenting campaign against NAFTA and bulk water export (www.canadians.org). Given Canadian attitudes to water, stronger legislation is expected from the federal level and in trade negotiations to promote sustainable water use and conservation. Concern remains about the definition of water under NAFTA, the effect of the investment and service provisions, and possible challenges to that agreement.[107] Two other topics of current and future interest are the increasing importance of the bottled water industry, given its acceptability as a means to export water from Canada; and the possibility that Canadian policies relating to water export could change.

Water in Bottles

Bottled water attracts attention despite its small volume compared to many bulk export proposals. It is an industry with both benefits and problems. Canada has seen growth in the export of unsweetened waters and ices, most activity occurring in the provinces of Ontario, BC and Québec. Canada's exports of bottled water increased from $130.5 million to $284.3 million between 1996 and 2002, but declined to $57.4 million in 2007 (Agriculture and Agri-Food Canada 2009). Competition with Canadian sources includes Evian from France and Fiji Water, bottled from a natural artesian aquifer at Yaqara, in northeast Viti in Fiji, a business started by Canadian David Gilmour. The switch to bottled water is driven by reductions in price, greater availability, more suppliers in the market, and more concern about obesity (Bloom 2005, B1, B6). Of course, events like the Walkerton poisonings are a factor as people come to mistrust the quality of their water supply or in disaster situations, are grateful for bottled water.

Bottled water is certainly fashionable. There are now international awards as for wine-tasting. At the Berkeley Springs International Water Tasting Competition in West Virginia in 2012, Jackson Springs, Manitoba won gold for best bottled water (Berkeley Springs 2012). Sommeliers can help you choose and enjoy your very expensive water: "You'll notice a foaminess on the palate, but not like Perrier, where it jumps out of your glass. The bubbles are very fine. It's balanced, not aggressive" (Pearce 2004, L4). More information for aficionados can be found in the archive of the *Water Connoisseur Newsletter* (fine waters 2009). Enjoy!

But there are downsides to all this water bottling activity, particularly when you take into account the plastic litter accumulating in areas not having recycling programs. Barlow and Clarke were among the first to raise concerns about wasteful plastic bottles and the price for bottled water (2003, 142-144). The Earth Policy Institute suggests that bottled water can cost 10,000 times more than tap water based on the cost of the plastic bottles (derived from crude oil) and their transport (Arnold and Larsen 2006). Second, although regulated as a food product under the federal *Food and Drug Act* and by a code adopted by some bottlers (Canadian Bottled Water Association 2009), the same quality standards as municipal water supplies may not be met (Eaton 2004, 64). In addition, exposure to chemicals such as *bisphenol A* leaching from hard plastic bottles is a serious health hazard that has caused stores to withdraw products in these containers. The general issue now means that water bottlers support recycling efforts and municipalities consider banning bottled water in public venues.

Bottlers are not paying enough for water, either because it comes from unlicensed sources such as groundwater or because many provinces do not charge anything other than a licence fee for this extraction purpose. Opposition to water-pricing proposals comes from the Bottled Water Association who feels that their share of actual water use is very small (in Ontario in 2002, 0.2 percent of all water taking permits) (Nowlan 2005, 72). Added to these concerns are fears about the localized effects of over extraction of groundwater; just ask the residents of Guelph, Ontario and surrounding areas (Fojtik 2007; Wellington Water Watchers 2007). Nestlé Waters Canada requested a 5 year renewal of its permit to extract 3.6 million litres of groundwater per day in Aberfoyle, a community near Guelph.

The Ontario government initiated a 30-day public review period and the response may have surprised them. Community residents, many inspired by the Wellington Water Watchers, overwhelmed the Ministry with 6,000 comments. The City of Guelph recommends in its Water Supply Master Plan that more sources for groundwater be found to supply an increasing population and that Nestlé's permit be renewed for only two years. But the Ministry renewed Nestlé's permit for 5 years. Guelph's situation requires provincial action: a source water protection plan in consultation with affected communities (now underway); legal tools for groundwater protection including local government land use controls; and appropriate pricing of groundwater withdrawals for bottling operations. This case is only one example of why the bottled water industry is raising concerns.

Will Water Export Policies Change?

Water export legislation or policies could change.[108] Water export could be proposed in return for access to markets or reductions in acid rain. A crisis from a regional economic decline or serious deterioration of water quality or humanitarian grounds would likely trigger water export. Technological innovations could result in the reduction of water diversion costs or adverse environmental effects. Governments could shift their policies from environmental protection to economic development giving credence to the comment: "There is no longer much doubt that the Government of Canada has, for the last two decades, looked at the water export issue very differently than the Canadian public. For Ottawa, it has not been about exchanging water for revenue, but using water as a lever to gain access for Canadian producers to the huge US market" (Quinn 2007, 13).

Just one of these drivers for lessening of water export control is climate change. Water will become more available in some areas, but more necessary for agricultural production in others, as the following example shows (Byers 2004, A13). In southern Alberta, agriculture depends on water from the St. Mary River, the North Milk River and the Milk River (which eventually flows back into the United States). In Montana, the St. Mary Canal, constructed in 1917, diverts water from the St. Mary River to the North Milk River for use by 17,000 northern Montana residents and irrigation of about 140,000 acres in the basin (Kline 2006). The rivers have their sources in Montana: the St. Mary from the high elevations of Glacier National Park in the summer and groundwater base flow in the winter, and the North Milk and Milk rivers from spring snowmelt and rainfall in the foothills of the eastern slopes of the Rocky Mountains. Climate change may cause the St. Mary River to have a more stable or increased flow if the glaciers that feed it melt more rapidly while the Milk River may decrease its spring flow due to decreased snow pack and increased evaporation. This may result in a short term advantage to Canada, but the longer term implications of climate change will likely bring serious consequences on both sides of the border (Byers 2004, A13). There are concerns even now.

Apportionment of the St. Mary and Milk Rivers is one of the disputes that led to the negotiation of the 1909 *Boundary Waters Treaty* and the file was reopened in 2003 by the IJC. The Commission received input from government and other sources and held public meetings, but has not been able to reach a satisfactory agreement. As of 2005, a Task Force will examine and report on measures to improve the existing situation (International Joint Commission 2005b).

Conclusion

Nova Group lawyer Robert MacRae said: "Canadian water has cachet. That is a tremendous component of the marketability of water" (cited in Scoffield 1998). Canada's water may have cachet, but many Canadians do not wish that advantage turned to water export. While such negative attitudes may be seen as an emotional nationalist response, why not? Water export brings together all the issues of water availability and quality protection. And there will always be the possibility of future conflict over water supplies despite a growing conservation ethos. With what result?

Aaron Wolf, an expert on water conflict, believes that "[w]ar over water is neither strategically rational, hydrographically effective, nor economically viable" (Wolf 1998, 261-262). He calls for cooperative water regimes established by treaty, "in advance of crises" (Nikiforuk 2004, 23). Echoing this, Peter Pearse, who chaired the *Inquiry on Federal Water Policy*, suggests that the furore masks what should be our real concerns: the need for better management, control over hydroelectric projects and over diversions from one watershed to another (Pearse 2000, A11). There is hope that any conflict will result in rational collaboration. The United States and Canada have a long history of bilateral relationships, particularly in the development of water projects and through the work of the IJC. Not only does the Commission make decisions, but it also acts as a facilitator for discussion and a repository for expert opinion (Doig 2002, 12). This continues in new forms as shown by the St. Mary and Milk Rivers Task Force.

Despite legislative and policy activity, more could be considered. Many people would like to see a new act to preserve Canada's water[109] and trade law clarification to clearly except water. Some ask why the federal government does not act under its "peace, order and good government" provisions (Saunders and Wenig 2007, 122). Government action relating to the bottled water industry could include enhanced licensing or permitting provisions, monitoring of local water supplies, appropriate pricing, and water quality standards. Maintaining current water export policies will require continued vigilance.

While the story of water export is not over, efforts to protect Canadian water should be applauded: public and interest group lobbying, press attention, policy and legislation. All are necessary to ensure sustainable water sources, as are information gathering about Canada's water demands and possible conservation measures. Environmental and social responsibility recognizes that: "water is a public trust; it belongs to the people. ... Exporting water for the elites who could afford it would reduce the urgency of finding real, sustainable and equitable solutions to water problems in the developing world. ... Neither poor people nor those trying to protect aquatic ecosystems have the financial resources necessary to compete for water in the open marketplace" (Barlow 1999, 6).

CHAPTER 11: PLANNING TOGETHER

Throughout this book we discover the challenges of water management. We need to consider water's quantity and quality together; manage groundwater and surface water as one; and allocate water accounting for demand, protection and conservation of the resource and the ecosystem. We must ensure good quality drinking water and protect development against flooding. It is essential to coordinate the efforts of various agencies working on one project or the multi-faceted issues presented by urban and resource uses. We need to accommodate physical (both natural and human-based aspects), economic, and social conditions as well as the political, legal, and institutional interests of many players. We will need to seek consensus, settle conflicts, and promote change.

"Planning" and/or "integrated management"[110] helps to resolve this complexity and uncertainty and achieve sustainability. Chapter 11 shows how these approaches have changed over time and now promise a better future through planning together for Canada's water. It uses examples of large-scale projects, basin-level planning, and strategic plans for water organizations; plans for water use, water allocation, water demand management or watershed management; as well as remediation, resource protection and disaster prevention plans. The chapter looks at planning efforts that have the potential to encourage sustainable water use. It tries to show how success "ultimately depends on the capacity of different actors and groups to communicate, negotiate and reach collective decisions" (Muro and Jeffrey 2008, 329). It concludes with a discussion of future directions.

Water Management Planning and Integrated Resource Management

What does it mean "to plan" in the context of water management? Definitions vary from "a logical and orderly way to think about the future (Dzurik 2003, 83)" to "water-centric planning": putting "water stewardship and sustainability front and center on the agenda of comprehensive land use, development, or resource planning initiatives" (waterbucket.ca 2006). The reality depends on what is planned for and by whom. A land use planner would say that planning means "planning of the scientific, aesthetic, and orderly disposition of land, resources, facilities and services with a view to securing the physical, economic and social efficiency, health and well-being of urban and rural communities." These activities include "conversion of land from natural habitats to urban built areas, maintenance and use of natural resources and habitats, and environmental protection" (Canadian Institute of Planners 2009).

Plans and zoning bylaws react to a combination of technological change, individual and developers' preferences, market and government failure, and imperfect information (e.g., time lag between cause and effect)(Zellner 2008, 441). Despite a demonstrated awareness of natural resource issues, the major focus is not on water. In planning for water, integrated resource management, "a process that aims for the sustainable use, management and development of water, land and related resources" (Capacity Building Network 2006) has become key. This approach strives to be "holistic and systems-oriented, [emphasizes] stakeholder involvement and partnerships, [views] water as an economic good, and [is] focused and goals-oriented" (Ramin 2004, 6). While integrated management may occur within or feed into land-use planning, the approaches may be two solitudes, a challenge to institutional adoption and belying the need for holistic consideration (Mitchell 2005).

The traditional process of planning includes gathering of information through research (relevant policy, legislation and regulations, programs, data, (and only sometimes ideas, needs, concerns and values, perceptions and beliefs)); analysis and synthesis contributing to the setting of principles, goals and objectives; design of the plan including reconciliation of competing objectives; passing of legislation if necessary, implementation of the plan's proposals, and then monitoring, evaluation and feedback into the next plan. The result has often been a master plan with a 5 year time horizon. But this emphasis on process and outcome may leave out the realities of political conflict and inadequately reflect the specific nature of the area being planned and the people planned for (Fainstein 2000, 452).

Today, many planners, and those who work in integrated resource management, define their roles not as creators of grand plans through a defined process, but as facilitators who listen and bring people's stories and ideas together with necessary technical information from many subject areas. There is also a growing awareness of the need for strategic change: "strategy-making is a process of deliberate paradigm change. It aims to change cultural conceptions, systems understanding and systems of meaning. It is more than just producing collective decisions. "It is about shifting and re-shaping *convictions*" (Healey 1997 cited in Friedmann et al. 2004, 51).

These processes confront organizational inertia to change that may be the result of complacency about the abundance of water, embedded relationships between clients and regulators, or simply lack of interest without the presence of a crisis (Heinmiller et al. 2008, 310).

As it relates to land or resource planning, strategic planning is distinguished from comprehensive planning in its attention to current means and circumstances rather than long-term trends, constraints and opportunities (Wiewel 2004, 61). The focus is on the creation of principles, goals, objectives, guidelines, and implementation (who will do what, when, where and how). It relates to key or selected variables (those causing the greatest variability and those most readily managed) (Mitchell 2005, 1338). The goal is to create a plan that will not "sit on a shelf". Consensus building and the tools of risk assessment and dispute resolution are essential.[111]

There are benefits and challenges to planning and integrated resource management as outlined by Andrew Dzurik in his book *Water Resource Planning* (now in a third edition). A plan gives direction to a series of decisions into the future, more predictable short-term decisions and a basis for monitoring and feedback. Both the public and agencies can participate, learn and share resources in a coordinated and cost-effective manner. Challenges may include time-consuming processes, legal obstacles to change, lack of sustainable funding, insufficient or outdated tools or lack of expert information, government inertia given day-to-day pressures, long-standing conflicts between agencies relating to control issues, and public apathy (Dzurik 2003, 109-11).

With these admonitions and encouragements, the next section examines plans of various types and scales, all relating the themes of this book. These examples augment those previously discussed for groundwater protection (Hornby Island's Official Community Plan), restoration and protection (Clean Annapolis River Project, St. Lawrence Vision 2000 Plan, Great Lakes Water Quality Agreement plans such as for Hamilton Harbour), disaster preparedness, and water supply (Guelph).

Types of Plans

This section begins with strategic planning for organizations, and then follows the book's thematic areas. Relating to water availability are water allocation plans, water demand management plans, and any plan providing quality protection by protecting water quantity. Public health and safety and aquatic ecosystem protection plans are for ecosystem restoration, drinking water quality, and floodplain areas. Plans for water development are exemplified by those for large projects (here dams, but historically for water export) and water use plans. Plans addressing multiple issues are at the basin or watershed level. Throughout this discussion, examples of the benefits and challenges of planning become apparent as does the way in which planning has changed and the importance of public process.

Strategic Planning for Water Organizations

Planning for water management organizations takes advantage of strategic planning techniques. While critiques of this process may suggest that it is all a public relations exercise or all about control (Mintzberg 1993, 32, 41), it is widely used to considerable effect. In some cases, particularly in the late 1980s and early 1990s, dramatic change came to older engineering-based groups as the growing importance of ecosystem-based planning was realized. As federal and provincial governments have down-sized their environment ministries and as others, of necessity or persuasion, take up the challenges, water-related organizations must plan strategically to ease change.

Strategic planning charts a future course to allow smooth transition from one initiative to the next (Mintzberg 1998, 15-17). These efforts encourage focus on key initiatives, reduce ambiguity, and provide order; all necessary where financial and human resources are constrained. A strategy also helps people define their own organization and distinguish it from others, recognize threats and opportunities, and emulate innovations. On the other hand, strategic planning may promise too much by trying to see far into the future, simplify a complex organization and hinder creativity (Mintzberg et al. 1998, 15-17). These are disadvantages to be aware of and to accommodate as part of monitoring and adaptive management efforts over the life of the plan.

How do water organizations develop a strategy for the future? My experience of strategic planning efforts is of organizations that built from the present to the future, but took into account accumulated wisdom and experience of many years. Besides seeking the advice of stakeholders, the knowledge of an organization's staff familiar with the water resource is crucial. We gain valuable insight; people come to understand their motivations better and become part of the process of change. Western Canada Water represents seven water and wastewater organizations from Alberta, Saskatchewan, Manitoba and the Northwest Territories. Their strategic plan's mission is: "Together we support and advance communication, education and training for water professionals in their efforts to provide safe water and to protect the environment" (*Western Canada Water* 2012). The plan states their values and proposes goals and objectives under the headings: Our People, Our Processes, Our Customers, Our Finances.

Water Availability Planning

Water Allocation Plans

While all protection plans affect water availability, those relating specifically to allocation (introduced in Chapter 2) are fundamental to many issues. Also related to water availability are water demand management plans; one type is "soft-path" plans.

At its most basic, water allocation measures out water to serve the multiple demands of various users. Provincial authorities carry out the process under a legislative framework and affect water rights held by a licensee or permittee. Historically, lack of technical information about the amount of water available and scientific knowledge about the effect of low flows hindered these processes. They have lacked staff trained to undertake overall assessments of available water supplies and systems to determine how to value the most important uses of water, be they environmental, social, cultural or economic. Also creating problems are "first-come, first-served" systems of allocation which issue licenses for lengthy terms, have no ability to transfer water from one use to another in some areas, and lack incentive to conserve water among users (New Zealand 2004). Water allocation planning adopts an integrated resource management approach and accommodates many of these issues, while allowing for adaptive management. Planning also reveals the 'political' aspects of decision-making and gives a better understanding of institutional and stakeholder interactions. It expands the range of expertise available beyond the traditional engineering and legal base. Ideally, it should take advantage of community knowledge beyond rights holders or water system operators.

The South Saskatchewan River Basin Plan[112] involves water allocation at a large scale, has been completed in two phases, and involved consultation with four multi-sector stakeholder advisory committees and the general public (Alberta Environment 2007). The first phase (approved in 2002) authorizes water allocation transfers within the basin, subject to approval by Alberta Environment, and establishes that no new applications will be accepted for the St. Mary, Belly and Waterton Rivers. The second phase (approved in 2006) establishes the status of water allocations and flows required to protect the aquatic environment, estimates future human demands for water as well as sub-basin flow contributions to a Master Agreement on Apportionment. The plan also guides decisions on licenses, preliminary certificates, approvals, or transfers of an allocation of water; supports development of water markets for transfers to accommodate re-distribution of water already allocated; and encourages improvements in water conservation by water users.

Water Demand Management Plans

One example of water demand management initiatives, "soft path" planning uses a strategic approach (Brandes and Brooks 2005). Taking a long term (20 to 50 years) view, the process works backwards to see how to get to the future. First, participants collect information about the capability of existing water services and population and economic projections. They estimate the quality and quantity of water required and identify new water sources, particularly those achievable through conservation. Then it is time to "back cast". Here Brandes and Brooks (2005, 13) ask why; for example, why do we use water to carry away our waste or use potable water in our gardens? Ultimately, they expect that the "soft path for water" will require behavioral change and many varied responses.

Protection Plans

Floodplain Management Strategies

Chapter 8 introduces ways to protect people and property against flooding. An overall strategy would take into account the complicated nature of these activities and the many organizations involved. Primarily associated with land use planning, floodplain management strategies seek a "resilient community" through individual and institutional actions that reduce risks and consequences (Halliday 2004, 105). Prepared in cooperation with every level of government, the plan would identify the flood hazard (extent and expected depth of flooding and how this affects existing land uses) and the policy for future land use. Possible structural controls such as dikes and guidelines for floodproofing can be carried forward in official community plans and zoning bylaws. These conditions are now commonplace, but there are other subjects that would ensure a more holistic approach: ecosystem protection, maintenance of and future needs for dams or dikes, financial responsibilities, and emergency response and recovery plans including mitigation and compensation policies. Plan preparation involving public consultation and education materials helps to raise awareness.

Discussion of the work of the Fraser Basin Council in Chapter 8 introduces the benefits of integrated floodplain management. The Ontario Conservation Authorities undertake similar work. But there is direct benefit to be gained from the preparation of local level plans. Okotoks, Alberta has a population of nearly 25,000 (2012) and the community was spared the severe damages of neighboring areas in Alberta's 2013 floods. The opening statement to their floodplain policy says:

> This policy document has been prepared ... to ensure that ... all policies of the Town of Okotoks related to development in flood prone areas:

- are readily available to all landowners and interested parties in a concise, comprehensive and clear document; and

- are integrated with other relevant planning policies related to the provision of services, appropriate densities and land uses, and the sensitive integration of *new development* in existing developed areas (Okotoks Planning Services 2001).

Prepared after public consultation, the Okotoks' document identifies the floodplain; municipal, provincial and federal responsibilities; and land use and development policies in the flood risk area. The Town's policy is that it does not intend to construct dikes and wishes to acquire the entire floodway for park purposes. The policy also provides servicing conditions for land in the flood fringe area where development is to be floodproofed. A brochure on the Town's website tells residents "What to do before, during & after a flood". Taken together, these documents provide a comprehensive strategy for protection.

Remediation and Restoration Plans

In 1988, subsequent to the 1978 Great Lakes Water Quality Agreement, Areas of Concern were designated for Remedial Action Plans in 43 locations around the Great Lakes (Kreutzwiser and de Loë 2004, 184). The focus of these plans is to bring stakeholders together and with good information, to arrive at feasible solutions (Slocombe in Mitchell 2004). Collingwood Harbour, discussed earlier, is exemplary of this approach.

The Niagara Peninsula Conservation Authority (1999) brings many different parties to the table in integrated resource management efforts with the goal of river restoration. This was exemplified by its strategy to restore the Welland River. This river has a drainage area of 880 square kilometres supporting agricultural, rural residential and urban development uses. Three regional municipalities and 11 local municipalities in the watershed area, including the City of Niagara Falls, need to deal with poor water quality, lack of diverse riparian/wetland habitat and physical barriers to fish migration. There is limited forest cover and a lack of good farm and non-farm management practices. The river has no proper and continuous drainage outlet and water levels fluctuate (Niagara Peninsula Conservation Authority 1999, 11).

The Welland River Restoration Committee guided the strategy's preparation. Prior to plan preparation, they undertook a review of needs and concerns of interested parties, discussion of technical and social issues in an open forum, the creation of communication links, and the appointment of a coordinating group who would also implement remediation methods. The strategy's simple but powerful goal: "To restore the ecological health of the Welland River and its watershed" sets specific objectives, with targets. For example, for the objective "[a] watershed that supports a healthy and diverse terrestrial ecosystem," the plan suggests that 30 percent of the land area should be natural forest and/or wetland. The plan succinctly describes the issue, the action to be taken, the project lead, and the partners in each action. In addition, and of utmost importance, the plan proposes costs ($3,250,000 over 10 years) and sets out what is to be done each year as well as the expected costs to each agency. Monitoring for both environmental and implementation targets is proposed. In 2009, the authority began updating the Strategy with a watershed plan.

British Columbia's Abbotsford Aquifer, a large unconfined aquifer, provides groundwater to over 100,000 people in Abbotsford and in Washington's Whatcom County. The aquifer has had serious contamination problems from nitrates since the early 1950s. Drinking water with high nitrate concentrations has the potential to cause a fatal condition called "blue baby syndrome" (Canada Environment 2004b). A planning process involved: information-gathering including delineation of the community wells' capture zone; documentation of potential contamination sources, risk assessment and development of ways to reduce pollution; and a contingency plan for the future (BC Health et al. 1995). Now an international task force coordinates management activities.

Water Development and Infrastructure Plans

Large-scale Projects

> "Crisis was to be turned into opportunity. We were the emissaries of hope."
> —Len Gertler on the Mactaquac Regional Development Plan of the 1960s to harness the St. John River (Hodge 1994, 40)

Large-scale water-related projects certainly are not as common as they once were. Typical of such single-purpose projects is the St. Lawrence Seaway, completed in 1959 at a cost of about $450 million, to allow ocean-going ships drawing less than 26 feet to travel to 15 ports on the Great Lakes.[113] Among the world's busiest shipping routes, the Seaway is once more subject to a planning process by Canadian and American officials as its infrastructure is ageing and it is unable to accommodate modern container ships. In 2002, in a draft report, the US Army Corps of Engineers proposed a $10-billion (U.S.) expansion that would deepen parts of the waterways (McKenna 2002, B1, B4). This ambitious plan raised many concerns. Environmental issues include possible lower lake levels, pollution (including invasive species), and erosion and threats to the Thousand Islands (Huard 2005, 3; McKenna 2002, B1, B4). A joint study, completed in 2007 sought ways to prevent operational degradation in the system over the next 50 years while optimizing existing infrastructure. Unlike the original Seaway project that would have followed a more traditional planning process, this study began with a series of stakeholder meetings and roundtable sessions (Huard 2005, 1).

Large hydroelectric projects also require significant planning, taking on all the characteristics of a regional planning exercise. Besides creating hydroelectric infrastructure, these projects can result in loss of communities and resources, as occurred in the Columbia River Basin (see Chapter 9). But "[e]xciting things happen when members of a community come together to create a shared vision of the future. They may even create magic as new light is shed on issues and new meaning is given to the way people see themselves and each other" (Beck 1997, 1). Believing that the downstream benefits from the Columbia River project should come back to the region most affected, the people of the Kootenays banded together in 1992[114] in a vision-driven planning process. Process and outcome were equally important. At meetings throughout the region and then at a symposium in Castlegar, delegates focussed on the benefits as well as the negative impacts of dam construction, reliving the past but also looking to the future of the region. The Columbia Basin Trust became the vehicle to guide that future, receiving a $295 million provincial endowment ($45 million for direct economic and business development). The remaining $250 million is to finance the construction of power projects. From 1996 to 2012, the Trust will also receive $2 million per year.

Today, the Columbia Basin Trust manages power and energy-related projects through a subsidiary company CBT Energy. The Trust works with Basin residents on water issues to increase understanding and build collective wisdom, acts as a facilitator and convener, and increases residents' opportunity for involvement and influence (Columbia Basin Trust 2007). These initiatives show how planning efforts have changed to reflect new thinking about how to plan, who should plan, and even what to plan.

Water Use Plans

BC Hydro and the BC government have created a process of "water use planning" for existing large reservoirs. An operating plan then guides system staff, taking into account stakeholder interests and natural conditions (McDaniels 1999, 501). BC Hydro undertook 24 plans in 5 years; their work began with principles for what should be planned for and the process:

1. The process must address multiple water use objectives: aquatic habitat, flood control, power generation (including costs and benefits of reservoir system changes), First Nations issues, land development, water supply, forestry, irrigation, navigation, recreation and cultural values.

2. Alteration of existing legal and constitutional rights and responsibilities are not part of the process, but legal rights are clarified.

3. Each process is to be site-specific, consider a range of alternatives, attempt consensus but allow for diversity of views, and be well-documented.

4. Trade-offs are to be recognized in monetary and non-monetary terms, recognizing that these occur within institutional constraints.

5. Consultative processes are to be inclusive; anecdotal, technical, and quantitative information is welcomed and is to be made available to all participants.

6. The plans are to be reviewed and revised (BC Environment, Land and Parks 1998).

Twenty to thirty stakeholders took part in each process: government agencies, First Nations, local communities, environmental and recreation groups and the public. Taking part meant embarking on an eight step integrated management process:

1. Initiate a Water Use Plan process for the particular facility

2. Scope water use issues and interests

3. Determine and initiate the consultation process

4. Confirm the issues and interests in terms of specific water use objectives

5. Gather additional information on the impacts of water flows on each objective

6. Create operating alternatives for regulating water use to meet different interests

7. Assess the trade-offs

8. Determine and document the areas of consensus and disagreement.

Following these steps, BC Hydro and government officials authorize the plan, then monitor compliance and review the plan (BC Hydro 2013).

British Columbia's water use plans help stakeholders make risk-related decisions in situations where consensus-building is difficult; stakeholders explain what matters to them, use the best technical information and recognize uncertainties, create attractive alternatives, identify trade-offs based on alternatives, and summarize areas of agreement and disagreement (Gregory 2001, 419). To do this, participants define "value-driven attributes". These are to be quantitative and qualitative; easily understood by stakeholders (e.g. when talking about floods, use the degree of safety achieved rather than the flood frequency); representative of the main parts of the system, but broken down into understandable components; and related to different geographic scales (if possible allowing small sections to be summed). The characteristics must be relevant to the policy alternatives under consideration; recognize money, time and personnel limits; and be open to change during the process (Gregory and Failing 2002, 493-495). Another technique is to use expert judgments about problem issues to replace data collection; a controversial technique, these opinions were thought to foster a more focused technical debate (Gregory and Failing 2002, 495). Finally, the process adopts a framework that would support adaptive management.

Water use plans do not come without significant costs to the power authority and government. These include plan process and implementation measures such as constraints on power generation to allow for flood control, the need for further studies, or the construction of new works. Costs can be recouped through a reduction in water rental fees paid by BC Hydro to the provincial government. Ultimately though, the cost is borne by the general taxpayer. Study costs of 20 draft water use plans were about $34 million per year, relative to licensed water rights, (intended to be amortized over 10 years), but this does not account for implementation costs (British Columbia Utilities Commission 2004a; 2004b).

The Province of Ontario has similar guidelines for planning the future of existing water power facilities.[115] Ontario's process is more detailed as it must deal with 83 waterpower producers. The plan proponent bears the principal costs while the Ministry of Natural Resources supplies some staff support and data collection costs (Ontario Natural Resources 2002, 8). Both BC and Ontario stress the need to determine the values of all users. Both provinces acknowledge the importance of information gathering. Both seek an open participatory process prior to the decision stage to assess options. Although said to be neither exhaustive nor prescriptive, the Ontario document contains specific information about consultation with First Nations (Ontario Natural Resources 2002, 47).

The Ontario principles highlight the need to reduce or eliminate adverse effects and increase in benefits without diminishing the financial performance of the water power operator. These principles direct that plans be made to stop degradation of the riverine ecosystem caused by the operation of water power facilities and, where possible, to restore the ecosystem. They recognize that the degree of impact on ecological processes is often uncertain and therefore will require monitoring and adaptive management. Planning is to be undertaken without prejudice to the rights of aboriginal peoples and treaty rights (Ontario Natural Resources 2002, 3, 14). Taken together, these examples from Ontario and British Columbia provide useful guidance to the creation of a process and principles for water management planning.

Multiple Issues (Basin to Watershed)

Ontario's 1946 *Conservation Authorities Act* introduced multi-purpose planning for flood control, wetlands conservation and water-based conservation at the watershed level within river basins (Hodge 1994, 38). With the *Canada Water Act* in 1970, the federal government supported large scale comprehensive water plans for BC's Okanagan Basin, New Brunswick's St. John River and the Qu'Appelle River, shared by Saskatchewan and Manitoba. Ultimately, high level support was lacking for these comprehensive plans. Their recommendations were planning rather than action-oriented, and typical for this period, the plan as a final product was seen as all important (Ramin 2004, 4). Today, planning at a basin-wide level is more likely to focus on overall strategic planning. While a basin-wide approach gives an understanding of the complex hydrology, geomorphology and ecological processes within a large system, a smaller scale (such as a watershed) highlights infrastructure developments and lessens the number of jurisdictions presenting institutional challenges (Sneddon 2002, 666).

The areal extent of a watershed[116] is promoted as an ideal base for water management planning because the physical boundaries offer a readily understandable focus. Watershed plans at the community level may combine various goals: protection of the aquatic ecosystem, drinking water of high quality and adequate quantity, managing water demand, and planning for the development and operation of infrastructure such as sanitary and storm sewers. An integrated watershed management approach combines planning for land and water use in urban, rural and resource use situations. It helps us to understand multiple users and their concerns and encourages sustainable use, but planning for watersheds can also be difficult (Table 11-1). Watershed plans, from the early work of Ontario's Conservation Authorities to more recent examples from other provinces, must have faced such problems yet have realized many benefits. Three examples show different purposes and diverse areas: a town and its watershed, a number of watershed areas where planning is by a provincial authority, and a watershed where responsibility is shared by a regional district and First Nations governance.

Newfoundland embarked on a watershed plan approach starting with the Gander Lake catchment area (Cantwell and Pinhey 1997). The lake supplies drinking water for the Town of Gander and other small municipalities. The rivers flowing into the lake support a rich salmon resource. But this important area is also the site of proposed forest harvesting, cottage development and a twinning of the Trans-Canada Highway. Early on in the planning process, non-point source pollution became an issue. The lake is very clean although sensitive to phosphorus loading; erosion transported by overland flow from soil caused by new development could be the culprit.

Table 11-1: Benefits and Problems of Watershed Planning
Sources: Adapted from Canadian Water Resources Association Manitoba Branch 2004, 27; France, R. (Ed.) 2005, 2-11; US Environmental Protection Agency 2003, 2

Benefits
Areal extent of watershed appropriate for planning area
Helps us to understand multiple uses and users
Scientific knowledge available at this scale
Many different organizations and types of staff can come together
Encourages partnerships and a multi-disciplinary approach towards options and common solutions
Provides a forum for public involvement, consensus building and capacity building
Encourages adaptive learning critical to managing watersheds
Problems
Lack of understanding of water management planning
Lack of resources—human and financial
Resistance to change a narrow focus
Lack of monitoring, enforcement, evaluation and feedback
Apathy / lack of ownership from stakeholders
Lack of a lead agency
Difficult to bring multiple jurisdictions and boundaries together
Lack of standardization for plans—what should be included
Weak legislation, licensing requirements, land ownership issues
Personal agendas (NIMBY) become embedded at beginning of process
Takes time for implementation
Data and technology capacity gaps
Government priority conflicts

A watershed decision model linked land use activity, geographic conditions such as soil and slope, and location within the catchment. Surprisingly, urban development and forestry activities were found to be less trouble than proposed cottage developments and twinning of the highway. The Final Management Strategy makes recommendations regarding future land use (based on potential to disturb the soil), buffer zones, new approaches for negotiating density of development, stormwater and cottage management guidelines, emergency response planning, and a water stewardship program. This plan provides an example of the importance of technical information in watershed planning to protect a valuable source of drinking water.

The Saskatchewan Watershed Authority is leading plan processes for watersheds and aquifers in 11 areas. Participants include rural, urban and aboriginal governments, Conservation Area Authorities, Watershed Associations and Irrigation District representatives. The plans may address just one issue or be comprehensive in nature (Saskatchewan Watershed Authority 2003). For example, the Lower Souris River Watershed Plan includes objectives and recommendations relating to many aspects: watershed education, groundwater threats and protection, community water supplies, water quality monitoring, water management and conservation. The plan affects many uses: landfills for waste disposal, municipal sewage lagoons, agricultural activities, fish migration and habitat, water supply for small communities, potential for failure of the Auburton Reservoir spillway, and operation of the Moosomin Dam (Saskatchewan Watershed Authority 2006, 8-22).

A Watershed Advisory Committee gives local input, guides the process, and shares in the implementation stage of the plan. A Technical Committee collects information and conducts analysis, where required, and the planning team (usually two persons from the watershed authority) coordinates activities and has the ultimate responsibility to develop the plan, but participation of other interest groups is also possible. Plans use a strategic approach. The resulting plan is a good model for clarity and use of implementation guidelines. Each of the issues identified includes an objective, recommendations, key actions, an implementation date, a completion date, a responsible entity, and for some recommendations, a success measurement. These plans will provide the basis for an adaptive resource management approach on the part of the Saskatchewan Watershed Authority.

Work on a plan for the Moberly Watershed near Fort St. John, BC is a final example; planning was for community-based watershed stewardship, rural residential water supply development, wastewater treatment, and community awareness and education. The process was a partnership of the West Moberly First Nations, Salteau First Nations, Peace River Watershed Council, Peace River Regional District and the Real Estate Foundation of British Columbia. Early findings of this process confirm the importance of several factors. Community-based watershed plans guide higher-level planning efforts and identify values and indicators understandable to community groups. A regional district can act as a facilitator working with senior government agencies and First Nations. The plan process relied on a blending of traditional ecological knowledge with contemporary science, early delineation of the community aquifer, linking ongoing outreach efforts to the process, and seeking consensus on watershed conditions as a basis for accommodating treaty resource-use rights (Whitten 2006, 22). Planning for the Bras d'Or Lakes in New Brunswick confirms the importance of this approach (brasdorcepi.ca c2006).

The watershed planning approaches described above have many benefits. The Gander Lake plan shows the importance of finding key information. The Lower Souris River Watershed Plan is particularly strong in the clarity and direction of its implementation sections. The Moberly watershed process provides not only the experience of planning between a regional district, senior governments and First Nations, but also some timely reminders about blending traditional ecological knowledge with contemporary science and making chosen values and indicators understandable to community groups. Communication is just as important as conflict resolution; plans act as an important overarching mechanism to achieve this (Margerum and Born 1995, 379, 383).

Future Directions

Water management planning will need to continue its focus on a combination of integrated management and land-use planning, to use strategic planning approaches and community capacity building, and to enhance monitoring. Given constrained economic conditions, innovation is the key to completion of time-consuming and costly planning processes.

Plans for drinking water quality protection have been given a high profile in recent years and from this, new challenges arise. Agencies frequently must plan for watersheds not totally within their jurisdiction; the purveyor needs to maintain water quality from source to tap and so becomes "real estate agent, land manager, regional economic development planner and environmental educator" (Pires 2004, 165, 173). Integrated watershed management needs to be combined with a multi-barrier approach to source water protection (FitzGibbon and Plummer 2004, 89). Implementation through land-use planning will be essential. Current legislative and institutional arrangements present barriers to this: "different values, views, languages, [and] practices", even different hydrological models. But there are shared opportunities in the possibility of local involvement and public education (Clay 2006, 2; FitzGibbon and Plummer 2004, 99-100).

Collaborative planning, building on stakeholder involvement, social learning, artistic expression and community capacity building, as described in Chapters 2 and 6, while not necessarily new, will benefit from more experience. For example, goals of the International Joint Commission's (IJC) watershed planning initiative are to use an ecosystem framework and to help local participants gain experience and build trust, while working together on problems (International Joint Commission 2005a, 1). For the St. Croix River watershed, the IJC's strategy is to prepare a comprehensive directory of interested organizations, develop a common vision for the watershed through transboundary digital watershed mapping and a hydrological model, and encourage "state of the watershed" reporting. The IJC hopes to act as an "honest broker" in the creation of partnerships between government agencies. Their role shows the continuing need for a government presence: three watershed strategies will cost $600,000 (U.S.) per year over a three-year period (IJC 2005a, 27, 28).

Key to the success of any plan is whether implementation and monitoring occurs and whether the results guide adaptive management. Some important factors include (Zbarsky 2005, 44-53): appropriately-funded plan administration, committed officials at all government levels, and staff who have the appropriate authority and training. Legal tools must be clear, consistent, and specify desired outcomes and non-compliance penalties, but these tools are not the only way of achieving goals.

To make implementation successful, a steering committee requires good facilitation, project leaders chosen as process drivers, as well as stakeholder involvement to build community ownership. In plan preparation, goals must be clear and priorities carefully chosen. After plan preparation, a business plan to get from strategy to operation is essential: "who, what, when, where and how" as a framework for each aspect that requires implementation" (Zbarsky 2005, 63).[117] Finally, careful monitoring, communication of results, and adaptation will provide a strong basis for on-going work of the plan and preparation of the next plan.

Many plans have been completed in recent years. We need to determine if their objectives were realized. Monitoring has traditionally been a matter of assessing the achievement of the plan. The people who develop the targets (performance indicators) often do the monitoring, but empowering people who have a vested interest to control the monitoring ensures action (Shordt 2000, 2-3). The range of measurement tools should be extended to include: water quality and quantity measures but also land changes, changes in land/water interfaces and biological indicators, the impact of the plan on people and their environment and perception changes such as "buy-in" and participation (Canadian Water Resources Association (Manitoba Branch) 2004, 29). Collecting on-going information allows immediate resolution of problems.

Conclusion

Previous chapters introduced water management as a subject of government, professional practice, business, academic and public interest and activity. On defining water management, it quickly becomes apparent that governance, a strategic approach, and strong leadership are keys in this complex subject area. What also becomes clear is the number of players involved and the interdisciplinary mix of their programs—be they operational, information-oriented, research-oriented, or activity-based. This chapter discusses planning and integrated resource management as an opportunity to bring all these players together. It shows processes to develop them, public participation to inform them, and ways to encourage successful plan preparation, implementation, and monitoring. At the start, the chapter recognizes the future orientation of planning and its attention to process. Planning has evolved over time so that now, strategic methods replace traditional planning. The change in process for the St. Lawrence Seaway revitalization and the Columbia Basin plans illustrate this evolution. In particular, work in both these areas now emphasizes stakeholder involvement and sustainability.

This chapter examines many different types of plans and attempts to summarize their contributions. Plans vary from strategic planning that helps organizations pursue a clear mandate within available resources to the Okotoks' plan that takes into account many organizations, a flood threat, and a developing community, as well as public perception of all of these. This chapter's examples illustrate important elements of planning processes. Gathering of information is usually the first need. Originally a step carried out by planners, now planners facilitate it in order to tap all sources. But the chapter gives examples of both. The plan to protect Gander Lake's drinking water and salmon resource was enhanced by science-based information gathering about key factors such as soil erosion. In the Welland River restoration plan, planners first reviewed concerns of interested parties, held open forums to discuss technical and social issues, created communications links and appointed a coordinating group to implement remediation.

Analysis and synthesis of information set the stage for innovation. BC Hydro's water use plans set out a framework within which all participants consider a range of alternatives, recognize trade-offs in monetary and non-monetary terms, attempt consensus but permit a diversity of views, and fully document the results. Goals determined for the plans discussed have a new, more ethical and holistic focus; for example, the Ontario process for hydro reservoirs starts from the point that degradation to riverine ecosystems should stop. When it comes to implementation, some plans have a legislative framework allowing: "clear, consistent directives ... sufficient jurisdiction over enough factors to enhance the likely attainment of the desired goals", but not without disadvantages: politically motivated agenda-setting and contradictory legislation of other authorities (Calbick et al. 2004, 48-49). To augment or replace the need for legislation, other plans provide detailed directions. The Lower Souris River Watershed Plan identifies each objective, key implementation actions, a completion date, who will be responsible, and for some recommendations, a success measurement.

The Moberly Watershed planning process brings together a number of important factors: use of community-based watershed plans to guide higher-level planning efforts, identification of values and indicators (traditional ecological knowledge with contemporary science) understandable to community groups, and a useful facilitation role of regional government.

Achieving sustainability means tackling what are known as "wicked" problems, given their complexity. Success occurs when all the players are at the table, working at an appropriate "political" and local level, and able to identify the important issues and understand the interests of others. Success depends on ensuring that technical procedures are combined with "sociological or pedagogical approaches"; people live in these places and there is a history (France 2005, 1-2).[118] Success means choosing the right kind of planning process. Water management planning gives the opportunity to recognize that each situation is unique, each requires new creativity, and everyone's ideas are important.

CHAPTER 12: CARING MORE

> We need to rediscover how to talk about change: how to imagine very different arrangements for ourselves ... (*Judt 2010, 53*).

Politicians and government bureaucrats have an essential part in caring for Canada's water, but ultimately each engaged member of society has a role. The present book recognizes and explains current water management approaches, seeking informed change, particularly from Canadian communities and individuals. *Canada's Water, Yours to Protect* acknowledges the world-wide recognition of a water crisis. Droughts, floods, water quality problems, over-allocation of water, and climate change translate into threats. These threats affect our health and the ecosystem, social customs and well-being, economic strength, and the intrinsic values accorded water. Reviewing how ideas have developed over time, both theory and practice, gives an introduction to the ways we now try to protect against the problems that beset water.

To know how the human use of water is managed is to recognize not just a professional undertaking, making top-down decisions, but more and more an activity promoting consensus and strategies among many. In Canada there are a host of players. All take part in this well-established area of interest; all their roles are necessary and are reflected in this book. Three chapters relate to the theme of water availability: water allocation, groundwater protection, and water demand management—efforts essential to protect the resource. The next three chapters focus on public health and safety and ecosystem protection; their effectiveness is a measure of sustainability. Two issues in the theme of water development and infrastructure, dam construction and water export, illustrate the challenges of such complex proposals. The necessity to resolve conflict and uncertainty related to water issues, to benefit from public and stakeholder involvement, and to arrive at sustainable solutions is demonstrated at every turn. Planning and integrated resource management tie together many of the ideas from previous chapters.

What Next?

Certainly the distance that water management has come since the early 1990s is impressive, as is the ingenuity displayed at realizing many of the solutions. Governments and communities are engaged. But there is no room for complacency. Current water sustainability efforts could be eroded by the reality of climate change (hydrological setting) and economic difficulties (human activities). Chapters 3 through 11 illustrate this and point to changes that seem likely to occur on a number of fronts. Future, more strategic, work is needed in three main areas:

- Governance: leadership, attention to the "political" aspects of water management, integration of land-use planning and resource management approaches, and more inclusive and better-informed decision-making;

- Social Responsibility and Commitment: awareness and engagement on water issues and through all water management actions by both actors and influencers; and an

- Enhanced "Water Ethic".

These suggestions appear at a time when making change is difficult. This does not mean that activity should cease until a more propitious moment. It is still essential to develop appropriate knowledge, policy and practice; to initiate smaller projects; and to be ready to seize opportunity when political leadership is available and/or in the presence of a crisis.

Governance

The federal government's role in water management continues to be a source of consternation, a perennial issue as that government's authority diminishes, given other priorities, and as provincial and local level activity increases. The Canadian national water policy adopted in 1987 is still considered to be the current policy. Should we look to the federal government to show leadership in the development of a future vision for water? Or should we look elsewhere and just accept that, at a senior government level, "by eviscerating the state's responsibilities and capacities, we have undermined its public standing (Judt 2010, 116)?"

It may be churlish to observe that taking responsibility for a vision over which you have little control may be fruitless. On the other hand, the Dublin Principles (1992) began a process that leads the dialogue about water even today. And such vision, at a national level, would hopefully increase political leadership and public awareness and guide funding priorities and the work of the 19 federal departments now having some responsibility for water. One example is the 55 Eco-Action projects funded between 2009 and 2012 across Canada (Canada Environment 2011b). Another, Agriculture and Agri-Food Canada's On Farm Action Program spent $85 million in the 2009-10 fiscal year for information for and implementation of Best Management Practices to cope with agricultural water shortages and climate change adaptation (Canada Treasury Board Secretariat 2010).

Many recent conferences and papers suggest what issues a national vision should address and how change could occur. The findings of this book concur with many of their suggestions (see endnote 1). Then there is the question of who else will show leadership. Academics and water professionals try, but this is a tall order with, in the case of the former, the need to carry a full teaching load, find research funding, manage graduate students, and "publish or perish". Water professionals in the private sector have clients to serve yet, like academics, make enormous efforts to spur dialogue. So do those who work in water management at government levels, but ultimately they are constrained by the need to not get out ahead of their political masters. Volunteer and interest-based organizations as well as the media have effected change. Each would have difficulty in assuming the role of overall leader, given their special interests and funding limits. It comes back to the federal government and to the political level. It is necessary to create a new and transparent process to realize a revived water vision for Canada (de Loë 2008; Muldoon and McClenaghan 2007, 259).

Provincial/territorial governments' strategy renewal has been active in the recent past and is continuing, much of it relating to drinking water quality. Their roles seem to lie most appropriately with the setting of vision, participation in inter-provincial initiatives, creation of legislation and standards, funding of local initiatives, and environmental assessment. Most important is support for information-gathering, planning efforts involving many local governments, and infrastructure. At the aboriginal government level, water governance models need development (such actions now lead by the Northwest Territories' *Northern Voices, Northern Waters: NWT Water Stewardship Strategy*).

There is a growing role for local and regional government in water issues. It is at this level that sustainability can be measured most effectively through a triple bottom line approach ensuring attention to economic, ecological and social parameters. Planning and integrated management highlight these factors and provide an ideal forum for community input (Chapter 11). For the future, "[p]lanning needs to promote local institutions ... that will engage citizens to learn and grow together through dialogue, will generate mutual trust and will provide the basis *for social trust, which in turn facilitates cooperative action*" (Laurian 2009, 385 (author's emphasis)). Better coordination between the various land and resource use agencies, even new institutional arrangements, may be necessary to encourage collaboration.

Three examples illustrate. First, provincial/territorial allocation of water continues, but changing approval authority to locally-based allocation plans is suggested. Second, adoption of the European Parliament's Water Framework Directive for watershed-based management, but a joint approach to the development of water quality standards is a goal worth pursuit (Bakker 2007, 365). Their model also applies this to floodplain management. Third, the wide-range of recommendations in current plans relating to drinking water is a tribute to the benefits of an integrated resource management approach, but this needs to extend to the specification of development limits in local land-use plans.

The next part of this issue is how to govern and with what information. Future research areas of interest include political aspects of water management and decision-making. Determining the appropriate roles involved in even one watershed can be difficult, but there is promise in studies adopting a political ecology approach (see Chapter 7). Equally important is to examine motivation (see Jane's story relating to Moncton's water in Chapter 2 and the floodplain management discussion (Chapter 8)) and how this affects decisions in specific situations. Malin Falkenmark, former Chair of the Scientific Program Committee for the Stockholm Water Symposium, and her colleagues understand this. They believe that for policy-makers, it is all about control, facilitation, and prioritization; for stakeholders, the need to identify and put forward their needs and interests; and for scientists and social scientists, the need to explain the links between water's characteristics and human requirements and desires, "the uncertainties in their statements, and the risk of non-decisions and stakes in making choices" (Falkenmark et al. 2004, 298). But it is also essential to be aware of motivation at the level of the elected official.

And we need to understand limitations suffered by water managers and policy-makers who often have all the technical skills necessary, but lack the necessary expertise to successfully speak truth to political power. Water management staff must have experience and qualifications to think beyond everyday solutions; this suggests the importance of interdisciplinary education/teamwork and knowledge of how to effectively welcome stakeholder and public input. Knowing how decision-making occurs (understanding context, settings, planning and appraisal activities, decision-making modes and actions) would promote less haphazard decisions and more clarity (Tonn et al. 2000, 163-171, 179-80).

To improve decision-making, we need better information from water observation networks, and better use and communication of that information. Knowledge about the supply and quality of surface and groundwater (basic hydrometric data, water quality parameters, expected flood flows, water's many values) is an obvious priority. Too often, funding for this vital function falls between the cracks, given other priorities. How to best communicate information to decision-makers, stakeholders, the media and the public is worthy of a study in itself.

Information technology is at the forefront of current water management activities. In the past, if writing about technology, you would immediately think of the eradication of water-borne disease, the effect of taking water from rural areas to supply large urban centres, and flood control efforts. Now, it's more likely you would consider new developments in information technology and data manipulation. The Internet has changed the way we do research and the speed at which this can be done. There are exciting ways to study water using search tools and online maps. There is huge potential in acquiring knowledge that can be translated into locally relevant guidance. Then good information becomes the basis for better decisions, the building of community capacity and planning processes.

Allied to all of this is government's need to choose priority action areas strategically, given reduced funding capability. Three general foci are suggested: groundwater information and management, water security ("access to adequate quantities of water, of acceptable quality, for human and environmental uses (Gordon Foundation 2006, 3))", and climate change. Chapter 4 describes the inadequacies of groundwater governance. As to water security, where the allocation systems in Canada fall short is not in recognizing the importance of the concept so much as in expanding aspects of its delivery. We need to look to planning approaches, innovative allocation and demand management tools (such as the use of water transfers and appropriate pricing), public input to decision-making, transparency in those processes, and monitoring of the results. Within all these actions, priority should be given to issues relating to aboriginal water rights. Guaranteed water quality, another aspect of water security, is now seen in the Northern Authorities Management Schemes and the Québec legislation, and will hopefully be considered by other areas. Adequate water quantity and quality for environmental purposes is the least easily defined area for action.

Emphasis on secure drinking water sources continues and is expected to do so unabated—this is the issue that has transfixed the media and the Canadian public—getting it right is essential. Simply reacting to new legislation is the first priority. Tied to this is the need for new and improved water supplies, difficult before, but now more so, given restrained economic times. Seeking funding for these improvements is essential: "Many of the infrastructure facilities in Canada will come to the end of their life in 10 to 20 years, ... We are proud of our ability to eliminate our fiscal deficit, but in the process we have what I call an infrastructure deficit of about $60-billion that is rising about $2-billion a year" (Won 2007, B8). It remains to be seen whether large-scale privatization of water and sewer services, perhaps part of the general trend to globalization of service provision, will seek to fill this gap. The potential is certainly there given ageing infrastructure that exceeds the capability of governments to finance new facilities, governments unwilling to raise taxes or charge fees sufficient to pay for infrastructure, and existing large private sector players able to enter the market.

The softer tools of water demand management are not sufficient here. The need to augment infrastructure is targeted by the deficit-funded economic stimulus package introduced to combat 2008's recession. Canada's Economic Action Plan allocated close to $2.5 billion for new water infrastructure over two years, but relied on a quick start-up for projects and matching grants from provincial and municipal governments. Even less encouraging, as these are grants and not loans, under-pricing of water is encouraged as the costs will not be passed directly to the benefitting taxpayer (Renzetti and Busby, 34).

For water management, the issue of climate change often brings with it a sense of hopelessness and fear. "Climate change is now not just a matter of debate over middleclass dinner tables, world summits of the eco-conscious and, perhaps, even (junior) government ministers or conferences of anorak clad scientists, who spend their lives dodging frostbite in the world's polar regions. It has become an issue about which it is virtually impossible to listen to any of the broadcast media without hearing mention or to pick up a newspaper or magazine without encountering an image of a receding glacier or the cracked and baked soil of a once fast flowing river bed" (Campbell 2006, 201).

These effects include more frequent floods and drainage problems, droughts, and associated water quality and ecosystem protection challenges. For the cause of each of these issues, the effect can be shown.[119] Rather than prevention, the question is how to adapt. Hydrologic models need to be linked with climate models, particularly at the regional scale (Morin and Cantin 2009, 3). Adaptation strategies for negative effects require understanding of climate change and disaster reduction together and financing options available at the local level.

Social Responsibility and Commitment

Throughout this book, I raise the importance of listening, really listening to those most affected: water licensees, owners of a well, those who speak for the protection of the ecosystem, consumers of drinking water, floodplain dwellers, people "in the way" of dam construction, and people affected by and/or contributing to any discussion about water policy and planning decisions. Frequently, attention to their voices has been missing in water management. Now their numbers increase and more strategic approaches to information gathering will be required if everyone is to speak and be understood. We need to know whether current government efforts to consult about policy change really make any difference. We need to understand that not everyone has sufficient stamina to pursue processes that can take many years.

Current efforts show the importance of social learning, building community capacity and other collaborative approaches. More experience will allow us to judge their utility. There are pitfalls: using these processes despite the presence of broader social processes or where communities are too poor or seriously conflicted, using participation as placation, and not recognising power disparities or the needed time or resources (Wismer and Mitchell 2005, 1). For the private sector, a study of natural resource sector players finds that they experience real opportunities to influence decision-making through collaborative processes, yet perspectives gained during the process do not significantly shape industry operations (Water Policy and Governance Group 2012, 5-6). There is a need to think of wider trade-offs and "no-regret" situations, to organize change in existing procedures, and to involve the public in long-term change (Tabara et al. 2005, 106). We want "a rebirth of imagination" (Mean 2007).

Allied both to governance and to creation of greater social awareness and engagement is the work of those who influence water management: the volunteer and private sectors, academic researchers, business, the media and the public. It is important to understand and expand their roles. For example, environment-related volunteer groups are only about 3 percent of the number of volunteer organizations in Canada, as compared to about 21 percent related to sports and 19 percent to religion (Murray 2006, 23). In a survey of 346 environmental stewardship organizations across Canada, the highest percentage (39 percent) was from groups that focus on ecological or habitat restoration, while 31 percent were involved in education (Lindsay 2006, 5). Environment groups receive 27 percent of their revenue from government sources (Murray 2006, 15). More research is needed relating to the water sector and how volunteer activities and funding can be most effective.

As to the private sector, Canadian companies' management of large projects in developing countries and privatization of water supplies at home are criticized, but the trend for companies to capitalize on their environmental knowledge, to show that sustainability is important to them, will hopefully continue. In a position paper prepared for the International Year of Freshwater (2003) and as a submission to the World Economic Forum, Alcan's (now Rio Tinto Alcan) spokesperson says: "While water is not our core business, without it, we would have no business at all. We want to contribute to preventing and resolving water crises by sharing our successes and the lessons we have learned in fresh water management throughout our 101-year history. No matter where we operate, it is our corporate responsibility to bring solutions to the table and to help manage the earth's water with care. Ensuring the sustainable management of water is essential to safeguarding our long-term license to operate and to grow" (Alcan 2003). To ensure continuation of the positive energy associated with the greening of the business community, government will need to place a strong emphasis on results-based management and transparency, and still encourage innovation, in order to retain public trust.

While it is always necessary to consult original government, interest group, and academic work, the media's reports bring forward emerging issues, keep these issues before the public eye, and follow-up on government action. There are academic and practical implications to this. We need to understand whether the media are acting as transmitters or transformers. As to the practical, government agencies, particularly water providers, can learn from special interest groups about how to harness public attention for positive change. Information can be made available to local communities on a consistent basis. "[P]eople generally respond more readily to carrots than sticks, ... they tend to avoid risk, and ... they act faster when they have easy access to clear information about how their behaviour compares with others" (Homer-Dixon 2009, A13). The media is an effective partner in such communications. Finally, we now recognize the power of "an engaged and mobilized public" who create "everything from pop stars to political history, proving that change is no longer the domain of a few well-connected people, but many people connected together" (controlcancer.ca 2009). These are the kinds of connections needed to protect water.

An Ethic of Care for Water

Water is no longer just another geological agent, but a moral issue (Tuan 1999, 106-7) that seeks responsibility in a framework of ethics and space, place, nature, knowledge and geography (Smith 1999, 272-288). The spatial extent of ethical concern can vary from, at a global level the human right to water to, at a local level, a watershed approach to planning by a community.[120] A concern for ethics and place means recognizing that places change through human action causing the need to prevent or mitigate the deleterious effects of waste disposal or water projects and balance "negative interdependencies"[121], understand the true costs of taking action, and pursue ethical investment. Considering ethics and nature means, first, recognizing water's basic functions as part of the ecosystem, giving life, creating landforms, providing landscapes and habitats, and transporting material and energy (Armstrong 2006, 13). And an ethic of care for water in nature means planning for sustainability, use of the precautionary principle during risk assessment, and ensuring environmental justice by promoting and protecting ecological integrity (Thiele 2000) and the intrinsic values of water.

These calls for an ethic of care for water have precursors, particularly in academic works and in the efforts of environmental pressure groups. But those seem insufficiently heard or sometimes, not heard at all. As challenges in water management become more complex and funding is targeted to other political priorities, a deeper sensibility is required. The need grows for renewal of ethical responsibility by all those who shape and influence water management.

In the activities described in this book, there are many times where ethical consideration requires enhancement. Show leadership at political and other levels. Demonstrate ethical awareness in the setting of aspirational values and principles and during practical decision-making; in the specification of guidelines for consultants, developers, and private-public partnerships; during consultation processes; in business ventures and in volunteer and public interest group efforts that pursue public good arguments. Opportunities exist in creating fair and well-researched media reporting, in academic studies, and in the ever-important intervention of individuals. Use collaborative community processes (social learning, community capacity building) in decision-making and seek a wider public to lead to the most equitable and efficient way to govern; encourage equity and transparency in water-related decisions. Make ethical investment decisions. Limit the development of unnecessary infrastructure. Understand and ensure water's true value. Infiltrate planning at every level with an ethic of care for water.

When teaching, use teamwork, encourage people to experience nature directly, provide grounding in community (including community struggles for social and environmental justice), study environmental ethics, and emphasize new literacies coming from alternative ways of being, knowing and acting/teaching (Sandercock 1998). When doing research, become an advocate, more involved in policy creation, and a solver of society's problems (Mitchell and Draper 1982). The question: "What shall it profit a profession, if it fabricates a nifty discipline about the world while that world and the human spirit are degraded (White 1972 in Platt 1997, 243)?" encapsulates this thought. When collecting information, listen, understand; use qualitative methods in research. Get close to reality by providing context, particularly historical background, bringing the "theories of justice and the human good, however tentative, to the facts of the local situation" (Smith 1999, 284, 288). When learning about the natural world, adopt a "lifelong syllabus—learning about ourselves in solitude, about our world in silence, and about our mutual relationship in some form of ritual- [enabling] our survival by being true *environmental stewards* (France 2006, 20).

Ethical responsibility extends to very specific actions: ensure protection of water's quality and quantity—speak first for the ecosystem and for water's own sake. Preserve floodplain ecosystems and allow no adverse impact through construction on floodplains. Delve deeper into the relationship between home and landscape and ask whether dams must be built. Recognize aboriginal peoples' water entitlements, mitigate the impact of dam construction on aboriginal territory, and protect aboriginal settlements on floodplain lands. Provide high quality water for every Canadian. These are all part of an ethic of care for water. Borrowing from the urban sphere, we need to seek recognition that water is part of our shared territory, values, public realm, support structures and destiny (Blakely and Snyder 1999, 44).

Conclusion

While there is still much to do, there is much to celebrate. The challenges abound, but the future is guided by knowledge and humility. One of 500 participants in the Peace Child International's world network, Canadian Connor Youngerman, writes (United Nations Development Program 2006, 17): "I imagined what it would be like—all this work, everyday, just for every sip of water. As we passed the cows, I noticed their water bin was a bit dirty and would need to be cleaned soon. The water that I get out of my tap is clean and clear and ready for drinking. But what if it came from a muddy river, or a stagnant pond? ... Why should I be entitled to this wealth and luxury? Why should I be water-fat, and others thirsty? Why do so many people need to worry where their next drink of water will come from? What can I do?" Canada's water is yours: care more, contribute to change, share what you learn.

BIBLIOGRAPHY

(Unless otherwise noted, website addresses are as of May 31, 2013.)

Abbott, R. 2009. *Conscious Endeavours: Essays on Business, Society and the Journey to Sustainability.* Scriptorium/Palimpsest

Acreman, M. 2001. "Ethical Aspects of Water and Ecosystems," *Water Policy.* 3: 257-265

Adams, A. 1984. *Ansel Adams: Letters and Images, 1916-1984.* Boston: Little, Brown

Agnew, C. and E. Anderson. 1992. *Water Resources in the Arid Realm.* New York: Routledge

Alberta Environment. 2007. *Approved Water Management Plan for the South Saskatchewan River Basin,* <http://environment.alberta.ca/1725.html>

Alberta Water Council. 2009. *Water Allocation Transfer Program Upgrade Project,* <http://www.awchome.ca/Projects/WATSUP/tabid/107/Default.aspx>

Alberta Wilderness Association. 2002. "Meridian Dam is Dead," *Alberta Wilderness Association Newsletter,* 11 March, <http://albertawilderness.ca/news/1999-2008/2002-awa-news-compilation>

Alcan. 2003. *Committed to the Sustainable Management of Water, One of Our Most Precious Resources,* 15 June 2007, <http://www.alcan.com/web/publishing.nsf/attachmentsbytitle/Sustainability-Docs/$-file/WaterPaper_F_AN.pdf>

Allen, D. and G. Matsuo. 2002. *Results of the Groundwater Geochemistry Study on Hornby Island, British Columbia. Final Report.* Burnaby, BC: Simon Fraser University. Earth Sciences, <http://www.islandstrust.bc.ca/poi/pdf/itpoitasrptgrndwtrfinalapr2002.pdf>

American Rivers. 1999. *Dam Removal Success Stories: Restoring Rivers through Selective Removal of Dams that Don't Make Sense,* <http://www.americanrivers.org/newsroom/resources/dam-removal-success-stories.html>

American Rivers and Trout Unlimited. 2005. *Exploring Dam Removal. A Decision Making Guide,* <http://www.americanrivers.org/newsroom/resources/exploring-dam-removal.html>

Anderson, T. and P. Snyder. 1997. *Water Markets: Priming the Invisible Pump.* Washington, DC: Cato Institute

Bibliography

Anderssen, E. 2009. "Fallen Hero: The Seaway at 50," *The Globe and Mail*. June 27

Annear, T. et al. 2009. *A Status Report of State and Provincial Fish and Wildlife Agency Instream Flow Activities and Strategies for the Future. Final Report for Multi-State Conservation Grant Project WY M-7-T*, <http://www.instreamflowcouncil.org/node/66>

Armstrong, A. 2006. "Ethical Issues in Water Use and Sustainability," *Area*. 38/1: 9-15

Armstrong, C. et al. 2009. *The River Returns*. Montreal: McGill-Queen's

Arnold, E. and J. Larsen. 2006. *Bottled Water: Pouring Resources Down the Drain*, <http://www.earth-policy.org/index.php?/plan_b_updates/2006/update51>

Arnstein, S. 1969. "A Ladder of Citizen Participation," *AIP Journal*. 35: 216-224

Association of State Floodplain Managers. 2008. *NAI-No Adverse Impact Floodplain Management*, <http://www.floods.org/NoAdverseImpact/NAI_White_Paper.pdf>

Atkinson, C. 2008. "Anger Grows Over Proposed River Project," *The Globe and Mail*. March 06

Atwood, M. 2000. "Looking Backwards, from 3000," *The Globe and Mail*. January 01

Audubon International. 2007. *Audubon Cooperative Sanctuary Program for Golf Courses*, <https://www.auduboninternational.org/Resources/Documents/ACSP%20Golf%20Fact%20Sheet.pdf>

Bakenova, S. 2008. "Making a Policy Problem of Water Export in Canada: 1960-2002," *Policy Studies Journal*. 36/2: 279-300

Bakker, K. 2003. *Good Governance in Restructuring Water Supply: A Handbook*, <http://powi.ca/wp-content/uploads/2012/12/Good-Governance-in-Restructuring-Water-Supply-A-Handbook-2003.pdf>

Bakker, K. (Ed.) 2007. *Eau Canada. The Future of Canada's Water*. Vancouver, BC: UBC

Ball, N. (Ed.). 1988. *Building Canada - A History of Public Works*. Toronto: University of Toronto. Précis, 02 November 2006, <http://www.infrastructure.gc.ca/research-recherche/rresul/hm/hm04_e.shtml>

Ball, P. 2001. *Life's Matrix: A Biography of Water*. Berkeley: University of California

Baltutis, J. and T. Shah. 2012. "Cross-Canada Check-up." Proceedings from the "Northern Voices, Southern Choices: Water Policy Lessons for Canada" *2011 National Discussion Series Tour: A Canadian Perspective on Our Water Future*, <http://poliswaterproject.org/publication/452>

Barlow, M. 1999. "Our Water's Not for Sale," *Canadian Perspectives*. Winter, 6

Barlow, M. and T. Clarke. 2003. *Blue Gold: The Fight to Stop the Corporate Theft of the World's Water*. Toronto: McClelland and Stewart

Bartlett, R. 1988. *Aboriginal Water Rights in Canada: A Study of Aboriginal Title to Water and Indian Water Rights*. Calgary: University of Calgary. Canadian Institute of Water Law

Beck, R. 1997. "Kootenay Lake Communities Create Their Own Future," Chapter 9 in *Sharing Stories*, 30 March 2012, <http://www.sfu.ca/cscd-new/gateway/sharing/chap9.htm>

Benidickson, J. 2002. *The Walkerton Inquiry Commissioned Paper 1—Water Supply and Sewage Infrastructure in Ontario, 1880-1990s: Legal and Institutional Aspects of Public Health and Environmental History*. Toronto: Queen's Printer for Ontario

Bibliography 193

—2007. *The Culture of Flushing: A Social and Legal History of Sewage.* Vancouver, BC: UBC

Berkeley Springs International Water Tasting. 2012. *Berkeley Springs International Water Tasting Awards,* <http://www.berkeleysprings.com/water/awards2.htm>

Biswas, A. 2005. "An Assessment of Future Global Water Issues," *International Journal of Water Resources Development.* 21/2: 229-237

Blaikie, B. 1999. "Motion re Fresh Water Resources," *Federal Hansard.* February 9: 11607-10177: 1015-40

Blakely, E. and M. Snyder. 1999. *Fortress America: Gated Communities in the United States.* Washington, DC: Brookings Institute.

Bloom, R. 2005. "Thirst for Water Takes Fizz Out of Cott Profit," *The Globe and Mail.* July 21

Bocking, S. 1998. Dams in Canada, 12 May 2007, <http://www.idsnet.org/Resources/Dams/Canadian/can-malaysia.html>

Booth, L. and F. Quinn. 1995. "Twenty-Five Years of the Canada Water Act," *Canadian Water Resources Journal.* 20/2: 65-90

Bourque, A. et al. 2007. *From Impacts to Adaptation: Canada in a Changing Climate,* <http://adaptation.nrcan.gc.ca/assess/2007/>

Boyd, D. 2003. *Unnatural Law. Rethinking Canadian Environmental Law and Policy.* Vancouver, BC: UBC

Boykoff, M. 2008. "The Cultural Politics of Climate Change Discourse in UK Tabloids," *Political Geography.* 27/5: 549-569

Brandes, O. and D. Brooks. 2005. *The Soft Path for Water in a Nutshell,* <http://poliswaterproject.org/publication/23>

Brandes, O. et al. 2005. *At a Watershed: Ecological Governance and Sustainable Water Management in Canada,* <http://www.poliswaterproject.org/publication/24>

Brandes, O. et al. 2010. *Worth Every Penny: A Primer on Conservation-Oriented Water Pricing,* <http://www.waterdsm.org/publication/344>

brasdorcepi.ca c2006. *The Spirit of the Lakes Speaks: Bras d'Or Lakes Collaborative Environmental Planning Initiative,* <http://brasdorcepi.ca/wp/wp-content/uploads/2011/07/Spirit-of-the-Lake-speaks-June-23.pdf>

Brinkman, M. (Ed.) 2003. "Cows and Fish: Alberta Riparian Habitat Management Program," *Enviroguide 2003.* 47

Briscoe, J. 1996. *Water as an Economic Good: The Idea and What It Means in Practice,* <http://rru.worldbank.org/Documents/PapersLinks/987.pdf>

British Columbia Drinking Water Review Panel. 2002. *Final Report: Panel Review of British Columbia's Drinking Water Protection Act,* <http://www.health.gov.bc.ca/protect/pdf/dwrp_final.pdf>

British Columbia Emergency Management. 2012. *The British Columbia Flood Response Plan,* <http://embc.gov.bc.ca/em/hazard_plans/BC-Flood_Response_Plan_2012.pdf>

British Columbia Environment. 2006. *Rights and Obligations of Water Licence Holders,* <http://www.env.gov.bc.ca/wsd/water_rights/water_rights.html>

—2009. *Annual Rental Rates for Water Licence Purposes by Sector*, <http://www.env.gov.bc.ca/wsd/water_rights/water_rental_rates/cabinet/new_rent_structure.pdf>

British Columbia Environment, Lands and Parks. 1991. *Sustaining the Water Resource*. Victoria: Environment, Lands and Parks Water Management Division.

—1993-1 - *Groundwater Management, Stewardship of the Water of British Columbia*. Victoria: Queen's Printer

—1993-2 - *Water Pricing, Stewardship of the Water of British Columbia*. Victoria: Queen's Printer

—1993-9 - *Background Report, Stewardship of the Water of British Columbia*. Victoria: Queen's Printer

—1993. *Stewardship of the Water Consultation*. Victoria: Environment, Lands and Parks Water Management Division

—1998. *Principles of Water Use Planning for BC Hydro*, <http://www.bchydro.com/content/dam/hydro/medialib/internet/documents/environment/pdf/wup_principles_of_water_use.pdf>

—2001. *Clean Water—It Starts with You—Agriculture*, <http://www.env.gov.bc.ca/wat/wq/brochures/agric.html>

British Columbia Environmental Protection Division. 2001. *Clean Water—It Starts with You—Pleasure Boating*, <http://www.env.gov.bc.ca/wat/wq/brochures/boating.html>

British Columbia Hansard. 27 March 1973. "Hon. Anderson Re Oak Hills Flooding," *Hansard*, <http://www.leg.bc.ca/hansard/30th2nd/30p_02s_730327p.htm>

—06 May 1975. "Hon. Anderson Re Oak Hills Flooding," *Hansard*, <http://www.leg.bc.ca/hansard/30th5th/30p_05s_750506z.htm>

British Columbia Health et al. 1995. *Fraser Valley Groundwater Monitoring Program*. Brochure

British Columbia Health Planning. Office of the Public Health Officer. 2001. *Provincial Health Officer's Annual Report 2000: Drinking Water Quality in British Columbia: The Public Health Perspective*, <http://www.health.gov.bc.ca/pho/pdf/phoannual2000.pdf>

BC Hydro. 2000. *Making the Connection. The BC Hydro Electric System and How It Operates*, 12 May 2007, <http://www.bchydro.com/rx_files/heritage/heritage6154.pdf>

—2006. *Appliance and Lighting Calculator*, <https://www3a.bchydro.com/appcalc/pg1.asp?id=0>

—2012a. *Peace River Generating Stations Project Update*, <http://www.bchydro.com/content/dam/hydro/medialib/internet/documents/projects/peace/A12_343_peace_river_ProjUpdates_2012_dec.pdf>

—2012b. *Revelstoke Unit 5 Project*, <http://www.bchydro.com/energy_in_bc/projects/revelstoke_unit_5.html>

—2013. *Common Questions. What are Water Use Plans?* <http://www.bchydro.com/about/sustainability/conservation/water_use_planning/questions.html>

British Columbia Land and Water BC. 2004. *Dealing with Drought. A Handbook for Water Suppliers in British Columbia*, <http://www.llbc.leg.bc.ca/public/pubdocs/bcdocs/371316/drought_handbook.pdf>

British Columbia Office of the Auditor-General. 1998/1999. *Protecting Drinking-Water Sources*, <http://www.bcauditor.com/pubs/1999/report5/protecting-drinking-water-sources>

British Columbia Office of the Auditor-General. 2010. *An Audit of the Management of Groundwater Resources in British Columbia*, <http://www.bcauditor.com/pubs/2010/report8/audit-management-groundwater-resources-british-columbia>

British Columbia Ombudsman. 2008. *Fit to Drink: Challenges in Providing Safe Drinking Water in*

British Columbia, <http://www.wsabc.ca/wp-content/uploads/2011/04/Ombudsmans-Report-on-Drinking-Water.pdf>

British Columbia Public Safety and Solicitor General. 2005. "Disaster Financial Assistance Limit Raised," *News Release*, <http://www2.news.gov.bc.ca/nrm_news_releases/2005PSSG0017-000325.pdf>

British Columbia Utilities Commission. 2004a. "5.0 Reference: Application, Volume I, Chapter 5, Heritage Contract," *Information Request No. 1.5.53*, <http://www.bcuc.com/Documents/Proceedings/BCH2004RR/CEC%20IR-1.pdf>

—2004b. *British Columbia Hydro and Power Authority. 2004/05 to 2005/06 Revenue Requirements Application and British Columbia Transmission Corporation Application for Deferral Accounts October 29*, <http://www.bcuc.com/Documents/Decisions/2004/DOC_5432_BCH%202004RR%20Final.pdf>

Brody, H. *Assessing the Project—Social Impacts and Large Dams*, <http://ideas.repec.org/p/ess/wpaper/id539.html#statistics>

Bronskill, J. 2003. "Terrorists May Poison Canadian Food or Water, Top-level Report Warns," *Canadian Press*, November 11, <http://circ.jmellon.com/docs/view.asp?id=520>

Brooks, D. 2006. "An Operational Definition of Water Demand Management," *International Journal of Water Resources Development*. 22/4: 521-528

Brooks, D. and R. Peters. 1988. *Water: The Potential for Demand Management in Canada*. Ottawa: Science Council of Canada

Brooymans, H. 2011. *Water in Canada. A Resource in Crisis*. Edmonton; Lone Pine

Brubaker, E. 2002. *Liquid Assets: Privatizing and Regulating Canada's Water Utilities*. Toronto: University of Toronto Centre for Public Management

Bruce, J. 2006. "Water Policy in Canada," *PRI Workshop*, 30 March 2012, <http://www.policyresearch.gc.ca/doclib/SD/PS_SD_Bruce(wed)_200605_e.pdf>

Bruce, J. and B. Mitchell. 1995. "Executive Summary," p. vi-ix in *Broadening Perspectives on Water Issues*. Ottawa: The Royal Society of Canada

Burby, R. et al. 1999. "Unleashing the Power of Planning to Create Disaster-Resistant Communities," *APA Journal*. 65/3: 247-259

Burke, B. 2001. *Don't Drink the Water: The Walkerton Tragedy*. Victoria, BC: Trafford

Burstyn, V. 2004. *Water Inc.* London: Verso

Buttimer, A. 1984. "Water Symbolism and the Search for Wholeness," pp. 159-290 in D. Seamon and R. Mugerauer (Eds.) 1984. *Dwelling, Place, and Environment*. Dordrecht: Nijhoff

Calbick, K. et al. 2004. "Watershed Resources Planning and Management: Lessons Learned from Comparative Case Studies," pp. 33-55 in D. Shrubsole. 2004. *Canadian Perspectives on Integrated Resources Management*. Cambridge, Ont.: Canadian Water Resources Association

Campbell, H. 2006. "Is the Issue of Climate Change Too Big for Spatial Planning?" *Planning Theory and Practice*. 77/2: 201-230

Campbell, K. and Y. Nizami. 2001. *Security or Scarcity? NAFTA, GATT and Canada's Freshwater*, <http://wcel.org/resources/publication/security-or-scarcity-nafta-gatt-and-canadas-freshwater>

Canada & the World Backgrounder. 2005. "Flooded Homeland," *Canada & the World Backgrounder*, 70/January Supplement: 41-42

Canada Mortgage and Housing Corporation. 1999. *Public-Private Partnerships in Municipal Infrastructure*, 12 November 2007, <http://www.cmhc-schl.gc.ca/en/inpr/su/waco/onrecast/onrecast_003.cfm>

——2000. *Household Guide to Water Efficiency*. Ottawa: Canada Mortgage and Housing Corporation

——2007 (accessed). *Train the Trainer Education Program*. EPCOR, Edmonton, Alberta, 29 April 2007, <http://www.cmhc-schl.gc.ca/en/inpr/su/waco/onrecast/onrecast_003.cfm>

Canada Agriculture and Agri-Food. 2003. *Groundwater*, 19 May 2003, <http://res2.agr.ca/publications/hw/02b3_e.htm>

——2006. *Water Supply and Irrigation for Agriculture*, 12 May 2007, <http://www.agr.gc.ca/pfra/water/supply_e.htm>

——2009. *The Canadian Bottled Water Industry*, <http://www4.agr.gc.ca/AAFC-AAC/display-afficher.do?id=1171644581795&lang=eng>

Canada Auditor General. 2011. *2011 June Status Report of the Auditor General of Canada*, <http://www.oag-bvg.gc.ca/internet/English/parl_oag_201106_04_e_35372.html#hd5f>

Canada Environment. 1985. *Currents of Change. Final Report*. Inquiry on Federal Water Policy. Ottawa: Environment Canada

——1987. *Federal Water Policy*. Ottawa: Environment Canada.

——1995. *Water: No Time to Waste - A Consumer's Guide to Water Conservation*, <http://www.ec.gc.ca/eau-water/default.asp?lang=en&n=344B115B-1>

——2003. *Threats to Sources of Drinking Water and Aquatic Ecosystem Health in Canada*, <http://www.ec.gc.ca/inre-nwri/default.asp?lang=En&n=235D11EB-1>

——2004a. *Almost Nine Million Canadians Depend on Groundwater*, <http://www.ec.gc.ca/eau-water/default.asp?lang=En&n=300688DC-1#sub5>

——2004b. *Freshwater Aquatic Ecosystems in Canada*, <http://www.ec.gc.ca/eau-water/default.asp?lang=En&n=6CA710A4-1#Section1>

——2004c. *The Properties of Water*, <http://www.ec.gc.ca/eau-water/default.asp?lang=En&n=BCCCF74B-1>

——2004d. *Water and the Canadian Identity*, <http://www.ec.gc.ca/eau-water/default.asp?lang=En&n=5593BDE0-1>

——2004e. *Water Quality Guidelines in Canada*, <http://www.env.gov.bc.ca/wat/wq/>

——2007. *Canadian Consumer Battery Baseline Study*, <http://www.ec.gc.ca/Publications/default.asp?lang=En&xml=C2F55D78-072A-4ED3-ACAA-DB4B4FE5B991>

——2010a. *Battery Recycling in Canada 2009 Update*, <http://www.ec.gc.ca/gdd-mw/default.asp?lang=en&n=52DF915F-1>

——2010b. *Frequently Asked Questions*, <http://www.ec.gc.ca/eau-water/default.asp?lang=En&n=1C100657-1>

——2010c. *Groundwater*, <http://www.ec.gc.ca/eau-water/default.asp?lang=En&n=300688DC-1#sub3>

——2010d. *Planning for a Sustainable Future: A Federal Sustainable Development Strategy for Canada*, <http://www.ec.gc.ca/dd-sd/default.asp?lang=En&n=4C1AB33B-1#s1>

—2011a. *Water Governance and Legislation*, <http://www.ec.gc.ca/eau-water/default.asp?lang=En&n=87922E3C-1>

—2011b. *Map of Eco-Action Funded Projects*, <http://maps-cartes.ec.gc.ca/ecogeo/Default.aspx?lang=en>

—2011c. *Water—Wise Water Use*, <http://www.ec.gc.ca/eau-water/default.asp?lang=En&n=F-25C70EC-1>

—2012. *A Renewed Commitment to Action: The 2012 Great Lakes Water Quality Agreement*, <http://www.ec.gc.ca/grandslacs-greatlakes/default.asp?lang=En&n=B274CBC1-1>

Canada Fisheries and Oceans. 1993/1994. *Stream Stewardship. A Guide for Planners and Developers*, <http://www.dfo-mpo.gc.ca/Library/189990.pdf>

Canada Foreign Affairs. 2009. *NAFTA—Chapter 11-Investment. Cases Filed against Government of Canada*, <http://www.international.gc.ca/trade-agreements-accords-commerciaux/topics-domaines/dispdiff/sunbelt.aspx?lang=eng>

Canada Health. 2008. *Boil Water Advisories and Boil Water Orders*, <http://www.hc-sc.gc.ca/ewh-semt/pubs/water-eau/boil-ebullition-eng.php>

Canada Industry. 2003. *What is a Public-Private Partnership?* 26 October 2006, <http://strategis.ic.gc.ca/epic/internet/inpupr-bdpr.nsf/en/h_qz01546e.html>

Canada Infrastructure. 2004a. *Assessing Canada's Infrastructure Needs: A Review of Key Studies*. Infrastructure Canada: Research and Analysis

—2004b. *Water Infrastructure: Research for Policy & Program Development*, 15 January 2011, <http://www.infrastructure.gc.ca/research-recherche/results-resultats/rs-rr/rs-rr-2004-01_01-eng.html>

—2006. *Adapting Infrastructure to Climate Change in Canada's Cities and Communities: A Literature Review*, 30 March 2012 <http://cbtadaptation.squarespace.com/storage/CdnInfrastructureAdaptation-LiteratureReview.pdf>

Canada Natural Resources. 2002. *Canada's Energy Markets*, 12 May 2007, <http://www2.nrcan.gc.ca/es/ener2000/online/html/chap3f_e.cfm>

—2004. *Climate Change Impacts and Adaptation: A Canadian Perspective*, <http://www.nrcan.gc.ca/earth-sciences/products-services/publications/climate-change/climate-change-impacts-adaptation/356>

—2012. *Distribution of Water*, <http://atlas.nrcan.gc.ca/site/english/maps/water.html>

Canada Policy Research Initiative. 2005. *Economic Instruments for Water Demand Management in an Integrated Water Resources Management Framework*, <http://www.obwb.ca/fileadmin/docs/economic_Instruments_water_demand_management.pdf>

Canada Public Safety. 2010. *Natural Hazards of Canada Map. Floods*, <http://www.publicsafety.gc.ca/res/em/nh/fl/index-eng.aspx>

—2011. *Canadian Disaster Database*, <http://www.publicsafety.gc.ca/prg/em/cdd/index-eng.aspx>

—2012. *Disaster Financial Assistance Arrangements DFAA*, <http://www.publicsafety.gc.ca/prg/em/dfaa/dfaa-guide-2008-eng.aspx>

Canada Statistics Canada. 2010. *Human Activity and the Environment. Freshwater Supply and Demand in Canada*, <http://www.statcan.gc.ca/pub/16-201-x/16-201-x2010000-eng.pdf>

Canada Treasury Board of Canada Secretariat. 2010. *Agriculture and Agri Food Canada—Report*, <http://www.tbs-sct.gc.ca/dpr-rmr/2009-2010/inst/AGR/agr00-eng.asp>

Bibliography

Canadian Bottled Water Association. 2009. *Model Water Bottling Code*, <http://www.cbwa.ca/resources/technical-training-tools>

Canadian Broadcasting Corporation. 2004. "Canada's Worst Ever E. coli Contamination," *CBC News*, December 20, <http://www.cbc.ca/news/background/walkerton/>

—2006a. "Move Northern Ontario Reserve South to Timmins, Says Adviser," *CBC News*, November 09, <http://www.cbc.ca/canada/story/2006/11/09/kashechewan-report.html>

—2006b. "North Dakota Revives 40-Year-Old Water Diversion Plan," *CBC News Manitoba*, January 11, <http://prairiedogguide.tripod.com/id16.html>

—2006c. "Water Treatment Methods," *News INDEPTH: Water*, September 26, <http://www.cbc.ca/news/background/water/treatment.html>

—2007a. "Alberta Announces Review of Water Transfers," *CBC News*, January 31, <http://www.cbc.ca/canada/calgary/story/2007/01/31/water-transfer.html>

—2007b. "SRB Technologies Loses Radioactivity Processing Licence," *CBC News*, January 31, <http://www.cbc.ca/technology/story/2007/01/31/srb.html>

—2010a. "Innu Warn of Global Fight against Hydro-Québec," *CBC News*, November 02, <http://www.cbc.ca/canada/montreal/story/2010/11/02/innu-protest-hydro-project-in-quebec.html>

—2010b. "Winnipeg's Water Rates Poised to Rise," *CBC News*, November 23, <http://www.cbc.ca/canada/manitoba/story/2010/11/23/mb-winnipeg-water-rate-hike.html>

—2011. "Manitoba Emergency Flood Channel Opens," *CBC News Manitoba*, November 01, <http://www.cbc.ca/news/canada/manitoba/story/2011/11/01/lake-manitoba-channel-opens.html>

—2012. *FAQs: Hydro-Fracking*, April 27, <http://www.cbc.ca/news/technology/story/2011/04/27/f-fracking-faq.html>

Canadian Council Ministers of the Environment. Water Use Efficiency Task Group. 1994. *National Action Plan to Encourage Municipal Water Use Efficiency*, <http://www.ec.gc.ca/Publications/default.asp?lang=En&xml=AD7BC046-AF7D-437A-BD39-A602B9042965>

—1999. *Accord for the Prohibition of Bulk Water Removal from Drainage Basins*, <http://www.ccme.ca/about/communiques/1999.html?item=13>

—2004. *From Source to Tap: Guidance to the Multi-Barrier Approach to Safe Drinking Water*, <http://www.hc-sc.gc.ca/ewh-semt/water-eau/drink-potab/multi-barrier/index_e.html>

—2006. *Water Conservation and Economics*, 24 February 2007, <http://www.ccme.ca/ourwork/water.html?category_id=84>

—2009. "Les Ministres de L'Environnement Collaborent sur le Sujet des Changements Climatiques," *Communiqués*. <http://www2.gnb.ca/content/gnb/fr/nouvelles/communique.2009.02.0180.html>

—2010a. *Review and Assessment of Canadian Groundwater Resources, Management, Current Research Mechanisms and Priorities*, <http://www.ccme.ca/assets/pdf/gw_phaseI_smry_en_1.1.pdf>

—2010b. *Water Valuation Guidance Document*, <http://www.ccme.ca/assets/pdf/water_valuation_en_1.0.pdf>

Canadian Dam Association. 2005. *CDA's Frequently Asked Questions*, 12 May 2007, <http://www.cda.ca/cda/main/faqdameffects.htm>

—2012. *Dams in Canada*, <http://www.imis100ca1.ca/cda/Main/Dams_in_Canada/CDA/Dams_In_Canada.aspx?hkey=63e199b2-d0e3-4eaf-b8ad-436d9415ad62>

Canadian Geographic. 2007. "Saguenay Flood," *Canada's Floods*, <http://www.canadiangeographic.ca/magazine/ma97/feature_saguenay_floods.asp>

Canadian Geosciences Council. 1993. *Task Force on Groundwater Resources Research Report. Groundwater Issues and Research in Canada*, <http://www.env.gov.bc.ca/wsd/plan_protect_sustain/groundwater/library/issues_research_Canada.html>

Canadian Heritage Rivers Network. 2001. *A Framework for the Natural Values of Canadian Heritage Rivers*. Second Edition, <http://www.chrs.ca/PDF/Natural_Values_e.pdf>

Canadian Institute for Environmental Law and Policy. 2004. *Public Participation in Water Management in the Great Lakes: Provincial and Joint Initiatives*. Toronto: Canadian Institute for Environmental Law and Policy, <http://www.cielap.org/pdf/GLPublicPart.pdf>

Canadian Institute of Planners. 2009. *Planning Is*, <http://www.cip-icu.ca/web/la/en/pa/3F-C2AFA9F72245C4B8D2E709990D58C3/template.asp>

Canadian International Development Agency. 2003. *Water Management*, <http://www.acdi-cida.gc.ca/INET/IMAGES.NSF/vLUImages/Performancereview4/$file/aaaEWLLpm.pdf>

Canadian Press. 2011. *Ontario Pays $72 Million to Victims of Walkerton Tainted Water Tragedy in 2000*, May 19

Canadian Water and Wastewater Association. 2004. *CWWA - Water Efficiency Experiences Database WEED*, <http://www.cwwa.ca/WEED/Index_e.asp>

—2012. *About the CWWA*, <http://www.cwwa.ca/about_e.asp>

Canadian Water Network. 2011. *Vision and Mission*, <http://www.cwn-rce.ca/>

Canadian Water Resources Association. Manitoba Branch. 2004. *Summary Report of the Integrated Watershed Planning and Management Forum*, 22 May 2007, <http://www.cwra.org/Manitoba_IW-PMF_Summary_Report.pdf>

Cantin, B. et al. 2005. "Using Economic Instruments for Water Demand Management: Introduction," *Canadian Water Resources Journal*. 301: 1—10

Cantwell, M. and J. Pinhey. 1997. "Watershed Management Plan for Gander Lake and Its Catchment," *Plan Canada*. 37/2: 27-30

Capacity Building Network for Integrated Resource Management. 2006. *Why Gender Matters. A Tutorial for Water Managers*, <http://www.unwater.org/downloads/why_gender_matters.pdf>

Capital Regional District. 1994. *Regional Source Control Bylaw*, <http://www.crd.bc.ca/wastewater/sourcecontrol/bylaw.htm>

Capital Regional District. Environmental Services. 2007. *Stormwater Codes of Practice and Best Management Practices*, <http://www.crd.bc.ca/watersheds/regulations.htm>

Carlson, H. 2004. "A Watershed of Words: Litigating and Negotiating Nature in Eastern James Bay, 1971-75," *Canadian Historical Review*. 85/1: 63-86

Carr, D. 2004. "Mobilizing Multi-Community Support for Heritage River Management by Facilitating Community Capacity Building," Paper presented at the 4th Canadian Heritage Rivers Conference: *Ribbons of Life: Sharing the Past—Charting the Future*, <http://www.grandriver.ca/River-ConferenceProceedings/CarrD.pdf>

Cavaye, J. 2000. *The Role of Government in Community Capacity Building.* Brisbane: Department of Primary Industries

Cech, T. 2003 *Principles of Water Resources.* U.S.: John Wiley

Chen, C. et al. (Eds.) 2013. *Thinking with Water.* Montreal: McGill-Queen's

Christensen, R. 2001. *Waterproof 2001: Canada's Drinking Water Report Card,* <http://www.ecojustice.ca/publications/reports/waterproof-canadas-first-national-drinking-water-report-card/attachment>

—2006. *Waterproof 2: Canada's Drinking Water Report Card,* <http://www.ecojustice.ca/media-centre/media-release-files/waterproof.II.exec.summary.pdf/view>

Christensen, R. and D. Droitsch. 2008. *Fight to the Last Drop. A Glimpse into Alberta's Water Future,* <http://www.ecojustice.ca/publications/reports/fight-to-the-last-drop-a-glimpse-into-alberta2019s-water-future>

Christensen, R. and S. Magwood. 2005. *Groundwater Pricing Policies in Canada,* <http://www.ecojustice.ca/publications/reports/report-buried-treasure-groundwater-permitting-pricing-in-canada/?searchterm=Buried Treasure>

Clark, M. 1990. "Water, Private Rights and the Use of Regulation: Riparian Rights of Use in British Columbia, 1892-1939," *The Advocate.* 48: 253-264

Clarke, B. 2010. 'In Victoria, Less Means More Means Less," *Globe and Mail,* October 23

Clay, R. et al. 2006. "Incorporating Drinking Water Source Protection into Watershed Planning: Thinking like a Sheep," Presentation to the Canadian Water Resources Association Conference *Working from the Source towards Sustainable Water Management,* 10 July 2006, <http://www.allsetinc.com/cwra2/files/presentations/2B_Robert_ Clay.pdf>

Clean Annapolis River Project. 2008. *Fast Facts and Notable Achievements,* <http://annapolisriver.ca/fastfacts.php>

Cochrane, Town of. 2006. *Town of Cochrane Water Conservation Based Water Rate Structure For Residential Accounts,* <http://www.cochrane.ca/municipal/cochrane/cochrane-website.nsf/AllDoc/B7ABDEAC954EB6B78725715D00659D30/$File/Water%20Conservation%20Meter%20Rate%20Report%20.pdf>

Columbia Basin Trust. 2007. *Water Initiatives Strategy,* <http://www.cbt.org/uploads/pdf/CBTWaterIntiativesFinalFeb2007.pdf>

Commission for Environmental Cooperation. Expert Workshop on Freshwater in North America. 2002. *Groundwater: A North American Resource. A Discussion Paper,* <http://www.cec.org/Storage/45/3781_water_disucssion-e1.pdf>

Conservation Ontario. 2011. *About Us,* <http://conservation-ontario.on.ca/about/index.html>

Conference Board of Canada. 2007. *Navigating the Shoals: Assessing Water Governance and Management in Canada,* <http://www.conferenceboard.ca/documents.asp?rnext=1993>

Constable, G. Papers held in the collection of the British Columbia Archives, Microfilm MSS 1462/A667, vol. 10, 11

Controlcancer.ca 2009. *The People vs. Cancer: Round 1,* 30 March 2012, <http://gopublic.squarespace.com/storage/pdf/wcd_ad_print.pdf>

Bibliography 201

Coote, D. and L. Gregorich. 2000. *The Health of Our Water: Toward Sustainable Agriculture in Canada*, <http://dsp-psd.pwgsc.gc.ca/Collection/A15-2020-2000E.pdf>

Council of Canadian Academies. 2009. *The Sustainable Management of Groundwater in Canada*, <http://www.scienceadvice.ca/en/assessments/completed/groundwater.aspx>

Council of Great Lakes Governors. 2001. *The Great Lakes Charter Annex. A Supplementary Agreement to the Great Lakes Charter*, <http://www.cglg.org/projects/water/docs/GreatLakesCharterAnnex.pdf>

Côté, F. 2004. *Freshwater Management in Canada: I. Jurisdiction*, <http://dsp-psd.pwgsc.gc.ca/Collection-R/LoPBdP/PRB-e/PRB0448-e.pdf>

—2006. *Freshwater Management in Canada: IV. Groundwater*, <http://www2.parl.gc.ca/content/lop/researchpublications/prb0554-e.html>

Cronon, W. (Ed.) 1996. *Uncommon Ground: Rethinking the Human Place in Nature*. New York: W.W. Norton

Cullen, A. 2003. "Letter to the Editor," *The Globe and Mail*. July 18

Curtis, W. 2002. "The Iceberg Wars," *Atlantic Monthly*. 289/3: 76-78

Dakin, S. 2003. "There's More to Landscape than Meets the Eye: Towards Inclusive Landscape Assessment in Resource and Environmental Management," *Canadian Geographer*. 47/2: 185-200

Daily Commercial News and Construction Record. 2008. *James Smith Cree Nation Pursues Billion-Dollar Hydroelectric Dam Project*, <http://www.dcnonl.com/article/id27062>

Dalyell, T. 2003. "Westminster Diary," *ENR: Engineering News-Record*. 254/22: 13

D'Alieso, R. 2011. "Chrétien's Call to Canada: Don't Fear Water-Export Debate," *The Globe and Mail*. March 23

Dauncey, G. 2004. "Ten Ways to Protect Your Water," *Corporate Knights*. Special Water Issue, 6

Davies, J-M. and A. Mazumder. 2003. "Health and Environmental Policy Issues in Canada: The Role of Watershed Management in Sustaining Clean Drinking Water Quality and Surface Sources," *Journal of Environmental Management*. 68: 273-286

Day, J. 1999. "Planning for Floods in the Lower Fraser Basin, British Columbia: Toward an Integrated Approach," *Environments. A Journal of Interdisciplinary Studies*. 27/1: 49

de Loë, R. 1999. "Dam the News: Newspapers and the Oldman River Dam Project in Alberta," *Journal of Environmental Management*. 55/4: 219-237

—2000. "Floodplain Management in Canada: Overview and Prospects," *Canadian Geographer*. 44/4: 355-68.

—2008. *Toward a Canadian National Water Strategy. Final Report*, <http://www.cwra.org/images/DocumentsLibraryAndConferenceProceedings/cnws_report_final_2008_06_18.pdf>

de Loë, R. and R. Kreutzwiser. 2003. *Protecting Drinking Water Sources: A Canadian Perspective*. Presented at the 2003 Annual Meeting of the Canadian Association of Geographers, May 26-31, 2003, Victoria, British Columbia

de Loë, R. et al. 2007. *Water Allocation and Water Security in Canada: Initiating a Policy Dialogue for the 21st Century*, <http://gordonfoundation.ca/publication/353>

de Villiers, M. 1999. *Water: Our Most Precious Resource.* Toronto: Stoddart

DeGrace, A. 2005. *Treading Water.* Toronto: McArthur

Desbiens, C. 2004. "Producing North and South: A Political Geography of Hydro Development in Québec," *The Canadian Geographer.* 48/2: 101-118

Diduck, A. 2004. "Incorporating Participatory Approaches and Social Learning," pp. 497-527 in Mitchell, B. (Ed.) *Resource and Environmental Management.* Don Mills, Ont.: Oxford

Diebel, L. 2008. "Canada Foils UN Water Plan," *The Star,* April 02, <http://www.thestar.com/News/Canada/article/409003>

Dinesen, I. 1972. *Out of Africa.* New York: Vintage Books

Dmitrov, R. 2002. "Water, Conflict, and Security: A Conceptual Minefield," *Society & Natural Resources.* 15/8: 677-691

Doig, J. 2002. "Territorial Modernity and Public Space: Lessons from the Politics of Water Conflict along the US/Canadian Border," *Innovation.* 13/1: 11-22

Donnelly, C. et al. 2002. "Decommissioning of Dams: Issues and Controversies Associated with Dam Removals," *Canadian Dam Association Bulletin.* 13/1: 7-20, <http://www.cda.ca/cda_new_en/publications/bulletin/bulletin.html>

Dorcey, A. 2004. "Sustainability Governance: Surfing the Waves of Transformation," pp. 529-554 in Mitchell, B. (Ed.) 2004. *Resource and Environmental Management in Canada.* Don Mills: Oxford

Dorcey, A. (Ed.) 1991. *Perspectives on Sustainable Development in Water Management: Towards Agreement in the Fraser Basin.* Vancouver: University of British Columbia Westwater Research Centre

Driedger, S. and J. Eyles. 2003. "Different Frames, Different Fears: Communicating about Chlorinated Drinking Water and Cancer in the Canadian Media," *Social Science and Medicine.* 56: 1279-1293

Dublin Conference on Water and the Environment. 1992. *Guiding Principles,* <http://www.gwp.org/en/About-GWP/Vision-and-Mission/>

Dzurik, A. 2003. *Water Resources Planning.* Lanham: Rowman & Littlefield

Eaton, J. 2004. *Bottled Water—Environmental Issues and Other Concerns in Canada and Beyond,* 15 September 2009, <http://www.sierraclub.ca/atlantic/programs/economies/water/SCCBottled%20WaterMarch05FINALDRAFTB.pdf>

Ecojustice Canada. 2010. *Walkerton's Lessons Poorly Learned,* <http://www.ecojustice.ca/media-centre/press-releases/walkertons-lessons-poorly-learned>

Economist. 2002. "Power of the Cree: A New Treaty Settles a Long-Running Row," *The Economist.* 362/8261: 43

—2003a. "Damming Evidence," *The Economist.* 368/8333: 9-11

—2003b. "Liquid Assets," *The Economist.* 368/8333: 13-15

Editorial. 1999a. "It's All for One in Blocking the Export of Bulk Water," *The Globe and Mail.* December 02

—1999b. 'Weirdness about Water," *The Globe and Mail.* February 13

Educom International, Inc. 1992. *The Export of Water from the British Columbia Coast: The Carter Report.*

Victoria, BC: Environment, Lands and Parks

Evenden, M. 2004. *Fish versus Power*. Cambridge: Cambridge University

Fainstein, S. 2000. "New Directions in Planning Theory," Urban Affairs Review. 35/4: 451-478

Falkenmark, M. et al. 2004. "Towards Integrated Catchment Management: Increasing the Dialogue between Scientists, Policy-makers and Stakeholders," *Water Resources Development*. 20/3: 297—309

Federation of Canadian Municipalities. 2012. *Canadian Infrastructure Report Card*, <http://www.fcm.ca/Documents/reports/Canadian_Infrastructure_Report_Card_Highlights_EN.pdf>

Feldman, D. 1991. "The Great Plains Garrison Diversion Unit and the Search for an Environmental Ethic," *Policy Sciences*. 24: 41-64

Ferguson, R. 2004. "Saving Canada's Sickest Lake," *The Globe and Mail*. December 11

fine waters. 2009. *The Water Connoisseur*, <http://www.finewaters.com/Newsletter/The_Water_Connoisseur_Archive/default.asp>

FitzGibbon, J. and R. Plummer. 2004. "Drinking Water Source Protection: A Challenge for Integrated Watershed Management," pp. 89-103 in Shrubsole, D. (Ed.) 2004. *Canadian Perspectives on Integrated Water Resources Management*. Cambridge, Ont.: Canadian Water Resources Association

Fitzgibbon J. et al. 2009. "What Can We Learn from Exemplary Groundwater Protection Programs?" *Canadian Water Resources Journal*. 34/1: 61-77

Flyvbjerg, B. et al. *Megaprojects and Risk: An Anatomy of Ambition*. Cambridge: Cambridge University

Fojtik, S. 2007. "Citizens Fight Nestlé over Town Water," *Inside the Bottle. Rabble*, June 15, <http://www.valemount.org/water/WaterDocs/CitizensfightNestle.pdf>

Fortin, P. 2004. "The Canadian Situation," *International Water Power & Dam Construction*. 56/12: 35-36

France, R. 2006. *Introduction to Watershed Development: Understanding and Managing the Impacts of Sprawl*, Lanham, MD: Rowman & Littlefield

France, R. (Ed.) 2005. *Facilitating Watershed Management. Fostering Awareness and Stewardship*. Lanham, MD: Rowman & Littlefield

Francis, Chief K. 1992. "First They Came and Took Our Trees, Now They Want Our Water Too," p. 93-101 in J. Windsor (Ed.) 1992. *Water Export: Should Canada's Water Be for Sale?* Cambridge, Ont.: Canadian Water Resources Association

Fraser Basin Council. 2003. *Floodproofing Options for Historic Settlement Areas*. Vancouver, BC: Fraser Basin Council

—2005a. *Authorities Affecting Source Water. British Columbia: Research Paper*, 30 March 2012, <http://www.fraserbasin.bc.ca/publications/fbc_reports.html> (see <http://www.rethinkingwater.ca/>)

—2005b. *Floodplain Management*, 30 March 2012, <http://www.fraserbasin.bc.ca/programs/flood.html#history>

—2005c. *Flood Hazard Management Tools and Information Resources Project*, 30 March 2012, <http://www.fraserbasin.bc.ca/programs/documents/FloodTools-Letter.pdf>

—2006a. *Fraser River Debris Trap*, 30 March 2012, <http://www.fraserbasin.bc.ca/programs/basin_wide.html>

—2006b. *Lower Fraser River Hydraulic Model—Summary of Results*, <http://www.fraserbasin.bc.ca/water_flood_model.html>

—2008. *Flood Hazard Area Land Use Management*. <http://www.fraserbasin.bc.ca/_Library/Water/report_land_use_and_flood_review_2008.pdf>

Fraser Basin Management Board. 1994. *Review of the Fraser River Flood Control Program*. A Task Force Report to the Fraser Basin Management Board. Vancouver, BC: Fraser Basin Management Board

Friends of the Trent-Severn Waterway. 1996. *Welcome to the Website of the Trent-Severn Waterway*, <http://epe.lac-bac.gc.ca/100/200/301/ic/can_digital_collections/trent_severn-ef/main_e_i.htm>

Friedmann, J. et al. 2004. "Strategic Spatial Planning and the Longer Range," *Planning Theory and Practice*. 5/1: 49-67

Galloway, G. 2011. "Delays Hold Up Water Supply on Reserves", *The Globe and Mail*. November 01

Gamble, D. 1998. "The Grand Canal and the National Interest," *Canspiracy*, <http://www.canspiracy.8m.com/article3.htm>

Gingras, P. 2009. "*Northern Waters: A Realistic, Sustainable and Profitable Plan to Exploit Québec's Blue Gold*," Montréal: Montréal Economic Institute, <http://www.iedm.org/files/juillet09_en.pdf>

Glass, A. 1988. "When the Elements Strike Back," *Times-Colonist*. September 21

Glenn, J. 1999. *Once Upon an Oldman: Special Interest Politics and the Oldman River Dam*. Vancouver, BC: UBC

Globe and Mail. 2003. "Settlement Reached in Drinking Water Case," *The Globe and Mail*. November 27

—2005. "Flood Leaves Town in State of Emergency," *The Globe and Mail*. September 29

—2006a. "Alberta Flood Damage Costlier than Expected," *The Globe and Mail*. January 10

—2006b. "Most Walkerton Residents Report Good Health," *The Globe and Mail*. November 09

—2008. "Making a Splash," *The Globe and Mail*. May 03

Goodman, A. 1984. *Principles of Water Resources Planning*. Englewood Cliffs, New Jersey: Prentice Hall

Gordon Foundation. 2006. *Protecting Our Water Resources*. Walter and Duncan Gordon Foundation Grantee Symposium, 30 March 2012, <http://www.gordonfn.ca/resfiles/water_symposium_summary.pdf>

—2011. *Canada's Great Basin*, <http://www.gordonfoundation.ca/sites/default/files/publications/Canada's%20Great%20Basin%20-%20Web.pdf>

Gordon Water Group of Concerned Scientists and Citizens. 2007. *Changing the Flow: A Blueprint for Federal Action on Fresh Water*, <http://www.flowcanada.org/sites/default/files/ChangingtheFlow.pdf>

Grant, T. 2006. "Water Heats Up for Investors," *The Globe and Mail*. October 12

Green Cross International. 2000. *National Sovereignty and International Watercourses Panel*, <http://webworld.unesco.org/water/wwap/pccp/cd/pdf/background_documents/national_sovereignty%20_international_watercourses_2000.pdf>

Gregory, R. et al. 2001. "Decision Aiding, Not Dispute Resolution: Creating Insights through Structured Environmental Decisions," *Journal of Policy Analysis and Management*. 20/3: 415-432

Gregory, R. and L. Failing. 2002. "Using Decision Analysis to Encourage Sound Deliberation: Water Use Planning in British Columbia," *Journal of Policy Analysis and Management*. 21/3: 492-499

Gutman, P. 1994. "Involuntary Resettlement in Hydro Power Projects," *Annual Review Energy Environment*. 19: 189-210

Haig-Brown, R. 1964. *A Primer of Fly Fishing*. Toronto: Collins.

Hajim, C. 2005. "Iceberg Hunters," *CNN Money*, <http://money.cnn.com/magazines/fortune/fortune_archive/2005/11/14/8360721/index.htm>

Haladay, J. 1999. "Letter to the Editor," *The Globe and Mail*. February 17

Halifax Regional Water Commission. 2006. *Water Quality Management. Maintaining Water Quality. The Multiple Barrier Approach*,<http://www.halifax.ca/hrwc/WaterQualityManagement.html>

Halliday, R. 2004. "Integrated Floodplain Management and Integrated Water Resources Management," pp. 104-123 in D. Shrubsole (Ed.) 2004. *Canadian Perspectives on Integrated Resources Management*. Cambridge, Ont.: Canadian Water Resources Association

Hannah, L. 1987. *From Abundant to Scarce Water Resources: An Evaluation of the Water Allocation Process and the Place of Instream Uses in British Columbia*. Victoria: University of Victoria: Unpublished Master of Arts thesis

Harris, L. 2002. "Water and Conflict Geographies of the Southeast Anatolia Project," *Society & Natural Resources*. 15/8: 743-759

Hatfield, C. and G. Smith. 1985. "Instream Resource Values and Protection Needs in Canada," *Inquiry on Federal Water Policy Research Paper No. 22*. Ottawa: Environment Canada

Haughton, G. 1998. "Private Profits-Public Drought: The Creation of a Crisis in Water Management in West Yorkshire," *Transactions of the Institute of British Geographers*. NS 23: 419-433

Hawthorn, T. 2004. "William Clancy 1915-2003. The Court Jester of B.C. Politics," *The Globe and Mail*. February 27

Heinmiller, B. 2003. "Harmonization through Emulation: Canadian Federalism and Water Export Policy," *Canadian Public Administration*. 46/4: 495-513

Heinmiller, B. et al. 2008. "Conclusion: Institutions and Water Governance in Canada," pp. 308-331 in M. Sproule-Jones et al. 2008. *Canadian Water Politics: Conflicts and Institutions*. Montreal: McGill-Queen's

Hickey, S. and R. Kolthari. 2009. "Participation," pp. 82-89 in *International Encyclopedia of Human Geography*. Manchester, UK: Elsevier

Hill, C. et al. 2008. "Harmonization Versus Subsidiarity in Water Governance: A Survey of Water Governance and Legislation and Policies in the Provinces and Territories," pp. 315-332 in *Canadian Water Resources Journal*. 33/4: 315-332.

Hobsbawm, E. 1994. *The Age of Extremes*. London: Michael Joseph

Hodge, G. 1994. "Regional Planning: The Cinderella Discipline," *Plan Canada*. 75th Anniversary Edition: 35-49

Homer-Dixon, T. 2009. "The Enticement of Green Carrots," *The Globe and Mail*. August 08

Horbulyk, T. 1997. "Canada," p. 37-45 in Dinar, A. and A. Subramanian (Eds.) 1997. *Water Pricing Experiences: An International Perspective*. World Bank Technical Paper No. 386. Washington, D.C.: World Bank

—2005. "Markets, Policy and the Allocation of Water Resources Among Sectors: Constraints and Opportunities," *Canadian Water Resources Journal*. 30/1: 55-64

Hornby Island Local Trust Committee. 2002. *Hornby Island Official Community Plan Bylaw No. 104, 2002*, <http://www.islandstrust.bc.ca/ltc/ho/pdf/hobylbaseocp0104.pdf>

—2010. *Submission from the Hornby Island Local Trust Committee on Water Act Modernization*,<http://islandstrust.bc.ca/ltc/ho/pdf/holatnewswateracmar222010.pdf>

Hornby Island Pilot Project Committee. 1994. *The Hornby Island Pilot Project. Final Report*. Victoria: Queen's Printer

Howlett, K. 2007. "Lack of Funding Putting Ontario's Water at Risk," *The Globe and Mail*. June 02

Hrudey, S. and E. Hrudey. 2004. *Safe Drinking Water: Lessons from Recent Outbreaks in Affluent Nations*. London: IWA Publishing

Huard, J. 2005. "The Québec and Cleveland Roundtable Sessions: A Positive Conclusion to the Stakeholder Engagement Process," *Communiqué*. 2/3: 1-4, 15 January 2011, <http://www.glsls-study.com/Supporting%20documents/Newsletters(English)/Newsletter%20October%202005%20(EN).pdf>

Hume, M. 2005. "Toying with Water Supply is a Recipe for Disaster," *The Globe and Mail*. December 12

—2006. "Heartlands Political Leaders Reject Power Plan," *The Globe and Mail*. October 04

—2007. "Virtual View of Power Projects Renders a Jolting Reality Check," *The Globe and Mail*. May 28

—2008a. "Government Seizes Polluted Site for Clean-up," *The Globe and Mail*. September 12

—2008b. "Province to Cap Copper Mine Leaking into Tsolum River," *The Globe and Mail*. April 15

—2009. "Kibosh on Province's Clean-Energy Call Fuelled by Logic," *The Globe and Mail*. August 03

—2011a. "Why Turn off the Tap on Nile Creek's Hatchery Success?" *The Globe and Mail*. April 04

—2011b. "Suit Presses Alcan to Release Water," *The Globe and Mail*. September 30

—2012. "Native Band Pushes for Water Licensing Reform," The Globe and Mail. November 14

Hume, M. and J. Hunter. 2012. "Band Proposes Relief Facility after Floods Overturn Graveyards," *The Globe and Mail*. June 1

Hunter, J. 2010a. "BC Hydro's Challenge: How to Tame and Shame Consumers," *The Globe and Mail*. October 22

—2010b. "Site C Lays Groundwork for Energy Exports Surge," *The Globe and Mail*. April 21

—2011. "Power Project Penance," *The Globe and Mail*. January 10

—2013. "A River Runs Through It," *The Globe and Mail*. January 19

Hurley, A. and A. Nikiforuk. 2005. "Don't Drain on Our Parade," *The Globe and Mail*. July 29

Ibbitson, J. 2000. "Testimony Raises Troubling Questions about Canada's Water," *The Globe and Mail*. December 08

—2002. "Unquenchable," *The Globe and Mail*. June 15

Innes, L. 2010. "Water Rights and Reconciliation," *Canadian Water: Towards a New Strategy*. McGill Institute for the Study of Canada Conference. March 24,<http://www.mcgill.ca/water2010/>

Insurance Bureau of Canada. 1994. *Statement of Principles Regarding Insurance and Natural Hazards*, <http://www.tcim.ca/library/flood.htm>

International Commission on Large Dams. 2003. *Montreal 2003: Description of Study Tours*, 15 January 2011, <http://www.intellabase.com/icold/montreal2003/en/1_5_1.htm>

International Commission on Large Dams. 2012. "Is the Era of the World Commission on Dams Definitely Belonging to the Past?" *News*, 30 <http://www.icold-cigb.org/GB/News/news.asp?IDA=254>

International Joint Commission. 2000. *Protection of the Great Lakes. Final Report to the Governments of Canada and the United States*. Ottawa: International Joint Commission

—2004. *Protection of the Great Lakes. Final Report to the Governments of Canada and the United States. Review of the Recommendations in the February 2000 Report*. Ottawa: International Joint Commission

—2005a. *A Discussion Paper on the International Joint Commission's Watersheds Initiative*, <http://www.ijc.org/php/publications/pdf/ID1582.pdf>

—2005b. *Accredited Officers of the St. Mary-Milk Rivers. Public Consultation Background Documents*, <http://www.ijc.org/rel/boards/smmr/public_consult-e.htm>

—2006. *News Release: IJC Releases Update on Its 1998 Report, Unsafe Dams?* <http://www.ijc.org/rel/news/060724_e.htm>

International Water Power & Dam Construction. 2004. "Mini-Hydro Comes On Line in BC," *International Water Power & Dam Construction*. 56/12: 4

—2005. "Blue Planet Prize Winners Announced," *Water Power*, <http://www.waterpowermagazine.com/news/newsblue-planet-prize-winners-announced/>

Irvine, J. 2002. "Water Law in Canada," *Just Add Water Conference*, Saskatoon, 27 December 2002, <http://www.saskriverbasin.ca/Conference/2002/Presentations/irvine/Jurisdiction.pdf>

Islands Trust Fund. c1998. *How to Protect Land*, 01 December 2003, <http://www.islandstrustfund.bc.ca/>

Johansen, D. 2001. *Bulk Removals, Water Exports and the NAFTA*, <http://dsp-psd.pwgsc.gc.ca/Collection-R/LoPBdP/BP/prb0041-e.htm>

—2007. *Bulk Water Removals: Canadian Legislation*, <http://www2.parl.gc.ca/Content/LOP/ResearchPublications/prb0213-e.htm>

Johns, C. and K. Rasmussen. 2008. "Institutions for Water Resource Management in Canada," pp. 59-89 in Sproule-Jones et al. 2008. *Canadian Water Politics: Conflicts and Institutions*. Montreal: McGill-Queen's

Johnson, C. et al. c2004. *Crises as Catalysts for Adaptation: Human Response To Major Floods*. ESRC Environment And Human Behaviour RES-221-25-0037 Research Report. Enfield, Mddx: Flood Hazard Research Centre

Johnson, Jr., C. 1988. "Historical Review of Drinking Water," pp. 136-139 in D. Speidel et al. 1988. *Perspectives on Water: Uses and Misuses*. New York: Oxford

Johnston, R. 1983. *Geography and Geographers: Anglo-American Human Geography since 1945*. London: Edward Arnold

Joseph, C. et al. 2008. "Implementation of Resource Management Plans: Identifying Keys to Success," *Journal of Environmental Management*. 88: 594—606

Judt, T. 2010. *Ill Fares the Land.* Penguin Group USA

Kahneman, D. and J. Knetsch. 1992. "Valuing Public Goods: The Purchase of Moral Satisfaction," *Environmental Economics and Management.* 22: 57-70

Kates, R. 2011. *Gilbert F. White. 1911-2006. A Biographical Memoir,* <http://nas.nasonline.org/site/DocServer/White_Gilbert.pdf?docID=74341>

Karvinen, W. and M. McAllister. 1994. *Rising to the Surface: Emerging Groundwater Policy Trends in Canada.* Kingston, Ontario: Queen's University Centre for Resource Studies

Kaya, A. "The Euphrates-Tigris Basin: An Overview and Opportunities for Cooperation under International Law." *Arid Lands Newsletter.* 44, <http://cals.arizona.edu/OALS/ALN/aln44/kaya.html>

Kelly, E. et al. 2010. "Oil Sands Development Contributes Elements Toxic at Low Concentrations to the Athabasca River and its Tributaries," *Proceedings of the National Academy of Sciences.* 107/37: 16178-16183

Kempton, K. 2005. *Bridge Over Troubled Waters: Canadian Law on Aboriginal and Treaty "Water" Rights, and the Great Lakes Annex,* <http://www.chiefs-of-ontario.org/node/96>

Kennet, S. 1991. *Managing Interjurisdictional Waters in Canada: A Constitutional Analysis.* Calgary: Canadian Institute of Resources Law

Kingwell, M. 2008. *Opening Gambits. Essays on Art and Philosophy.* Toronto: Key Porter.

Kinkhead, J. et al. 2005. *An Analysis of Canadian and Other Water Conservation and Practices,* <http://www.ccme.ca/assets/pdf/kinkead_fnl_rpt_2005_04_2.1_web.pdf>

Kline, L. 2006. "St. Mary Working Group Reaches Turning Point," *The Havre Daily News,* February 23, 15 January 2011, <http://www.havredailynews.com/articles/2006/02/23/local_headlines/stmary.txt>

Knafla, L. 2005. "Introduction: Laws and Societies in the Anglo-Canadian North-West, Frontier and Prairie Provinces, 1670-1940," pp. 1-55 in Pue, W. 2005. *Laws and Societies in the Canadian Prairie West,* 1670-1940. Vancouver: UBC, <http://www.ubcpress.ca/books/pdf/chapters/knafla.pdf>

Kneffel, E. 2002. *Hornby Island Groundwater Protection Pilot Project Phase II Report,* <http://www.islandstrust.bc.ca/poi/pdf/itpoitasrptgrndwtrmay2002.pdf>

Kohut, A. 1999. "Humans Interacting with Groundwater, Canadian Literary Images," pp. 429-437 in Alila, Y. (Ed.) 1998. *Mountains to Sea: Human Interaction with the Hydrologic Cycle.* CWRA 51st Annual Conference Proceedings. Cambridge, Ontario: Canadian Water Resources Association

Kohut, A. et al. 1993. Chapter 9—*Ground Water Resources of the Basins, Lowlands and Plains. 9.1.3 Gulf Islands,* <http://www.env.gov.bc.ca/wsd/plan_protect_sustain/groundwater/gwbc/C0913_Gulf_Islands.html>

Kretzmann, J. and J. McKnight. 1993. *Building Communities from the Inside Out: A Path toward Finding and Mobilizing a Community's Assets.* Evanston, Illinois: The Asset-Based Community Development Institute

Kreutzwiser, R. and R. de Loë. 2004. "Water Security," pp. 166-194 in Mitchell, B. (Ed.) 2004. *Resource and Environmental Management in Canada.* Don Mills, Ont.: Oxford

Kreutzwiser, R. et al. 2004. "Water Allocation and the Permit to Take Water Program in Ontario: Challenges and Opportunities," *Canadian Water Resources Journal*. 29/2: 135-146

Kumar, A. et al. 2001. *Managing Flood Hazard and Risk. Report of an Independent Expert Panel*, <http://dsp-psd.communication.gc.ca/Collection/D82-70-2002E.pdf>

Laidlaw, D. and M. Passelac-Ross. 2010. *Water Rights and Water Stewardship: What about Aboriginal Peoples?* <http://ablawg.ca/2010/07/08/water-rights-and-water-stewardship-what-about-aboriginal-peoples/>

Lantz, V. et al. 2010. *Valuing Wetlands in Southern Ontario's Credit River Watershed*. Mississauga, Ont.: Pembina Institute and Credit Valley Watershed.

LaForest, R. and M. Orsini. 2005. "Evidence-based Engagement in the Voluntary Sector: Lessons from Canada," *Social Policy & Administration*. 39/5: 481-497

Lake Ontario Waterkeepers. 2006. *Clean Water Primer #1. Taking Water from the Great Lakes: A Citizen's Guide to the Policies, Rules, and Procedures that Protect Ontario's Waterways*, <http://www.waterkeeper.ca/documents/primer1.pdf>

Lambert, S. 2009. "Manitoba to Spend Millions Repairing Flood Damage," *The Globe and Mail*. July 07.

Laurian, L. 2009. "Trust in Planning: Theoretical and Practical Considerations for Participatory and Deliberative Planning," *Planning Theory & Practice*. 10/3: 369—391

Layton, D. 2005. *The Bird Factory*. Toronto: McClelland & Stewart

Lebel, M. 2008. *The Capacity of Montreal Lake, Saskatchewan to Provide Safe Drinking Water*. Saskatoon: University of Saskatchewan Unpublished MA Thesis

Lewington, J. 2005. "Calgary Rain Close to Breaking Record," *The Globe and Mail*. June 20

Lindgren, R. 2003. *In the Wake of the Walkerton Tragedy: The 10 Top Questions*. Canadian Bar Association National Symposium on Water Law. Vancouver, BC: Canadian Bar Association, March 28, 29

Lindsay, L. 2006. *Family Volunteering in Environmental Stewardship Initiatives: Research Report*, <http://library.imaginecanada.ca/files/nonprofitscan/kdc-cdc/evergreen_family_volunteering.pdf>

Liu, K. and A. Baskaran. 2005. "Using Garden Roof Systems to Achieve Sustainable Building Envelopes," *Construction Technology Update*. 65, <http://archive.nrc-cnrc.gc.ca/eng/ibp/irc/ctus/ctus-n65.html>

Lucas, A. 1990. *Security of Title in Canadian Water Rights*. Calgary: Canadian Institute of Resources Law

Luciw, R. 2011. "The Basement Hell of High Water," *The Globe and Mail*. May 10.

Maas, T. 2003. What the Experts Think: *Understanding Urban Water Demand Management in Canada*, <http://poliswaterproject.org/publications>

MacDonnell, L. 2009. "Return to the River: Environmental Flow Policy in the United States and Canada," *Journal of the American Water Resources Association* (JAWRA). 45/5: 1987-1099.

MacGregor, R. 2009. "A Visionary's Epiphany about Water," *The Globe and Mail*. October 05

Mackie, R. 2004. "Manitoba-Ontario Hydro Project Moves Step Closer to Reality," *The Globe and Mail*. October 01

MacKinnon, M. 2001. "Grimes May OK Water Exports," *The Globe and Mail.* March 28

Mackrael, K. 2013. "First Nation, Winnipeg Battle over Water," *The Globe and Mail.* January 11

Mahan, R. et al. 2002. "Market Mechanisms and the Efficient Allocation of Surface Water Resources in Southern Alberta," *Socio-Economic Planning Sciences.* 36: 25-49

Makin, K. 2006a. "Charges against Ontario Utility Quashed," *The Globe and Mail.* November 15

—2006b. "Judge Acquits Supervisors in Dam Tragedy," *The Globe and Mail.* December 19

Margerum, R. and M. Born. 1995. "Integrated Environmental Management: Moving from Theory to Practice," *Journal of Environmental Planning and Management.* 38/3: 371-391

Matsui, K. 2005. "White Man Has No Right to Take Any of It" *Secwepemc Water-Rights Struggles in British Columbia,"* Wicazo Sa Review. 20/2: 75-101

Mayhew, S. 1997. *A Dictionary of Geography.* Oxford: Oxford University

McAllister, D. 1982. *Evaluation in Environmental Planning: Assessing Environmental, Social, Economic, and Political Trade-offs.* Cambridge, Mass.: MIT

McCarthy, S. 1999. "Ban on Water Exports Needed, Report Warns," *The Globe and Mail.* May 20

—2010a. "Churchill Hydro Deal Signals Era of Atlantic Co-operation," *The Globe and Mail.* November 19

—2010b. "Governments Vow to Overhaul Environmental Monitoring," *The Globe and Mail.* December 22

McCutcheon, S. 1991. *Silenced Rivers.* Montreal: Black Rose Books

McDaniels, T. 1993. "Contingent Valuation and Multiple Objective Approaches Compared," in BC Environment, Lands and Parks. *Full Cost Accounting & The Environment.* Seminar Proceedings. Victoria, B.C.: BC Environment, Lands and Parks

McDaniels, T. et al. 1999. "Democratizing Risk Management: Successful Public Involvement in Local Water Management Decisions," *Risk Analysis.* 19/3: 497-510

McDougall, M. 1999. *An Overview of Water Regulation in Saskatchewan.* Canadian Bar Association National Symposium on Water Law. Toronto, April

McKenna, B. 2002. "Seaway Idea Hits Wave of Resistance," *The Globe and Mail.* June 10

—2012. "How Sandy Revealed the Follies of Federal Flood Insurance," *The Globe and Mail.* November 12

McNeil, K. 1999. *Riparian Rights and Land Reserved for the Indians: Some Constitutional Issues.* Canadian Bar Association National Symposium on Water Law. Toronto, April

McNeill, R. 1991. "Water Pricing and Sustainable Development in the Fraser River Basin," pp. 417-429 in Dorcey, A. 1991. *Perspectives on Sustainable Development in Water Management.* Vancouver, BC: UBC Westwater Institute

Mean, M. (2007) "Running on Empty," *Guardian Unlimited.* May 23

Mickleburgh, R. 2005a. "Something in the Water," *The Globe and Mail.* November 25

—2005b. "Spill Leaves River 'Dead'," *The Globe and Mail.* August 08

—2006. "Vancouver Lifts Water Advisory," *The Globe and Mail.* November 28

Million, L. 1992. *"IT WAS HOME": A Phenomenology of Place and Involuntary Displacement as Illustrated by the Forced Dislocation of Five Southern Alberta Families in the Oldman River Dam Flood Area*. San Diego, California: Saybrook Institute, Unpublished Doctoral Dissertation

Milroy, S. 2006. "What's a Dead White Guy Doing in the Middle of Our Gallery?" *The Globe and Mail*. January 21

Minardi, J-F. 2010. "Going down the Tubes," *Fraser Forum*, <http://www.fraserinstitute.org/uploadedFiles/fraser-ca/Content/research-news/research/articles/going-down-the-tubes.pdf>

Mining Watch. 2005. *Acid Mine Drainage Sites in Canada*, <http://www.miningwatch.ca/en/acid-mine-drainage-sites-canada>

Mintzberg, H. 1993. "The Pitfalls of Strategic Planning," *California Management Review*. 36/1: 32-47

Mintzberg, H. et al. 1998. *Strategy Safari. A Guided Tour through the Wilds of Strategic Management*. New York: Free Press

Mitchell, B. 1984. "The Value of Water as a Commodity," *Canadian Water Resources Journal*. 9/2: 30-37

—2005. "Integrated Water Resource Management, Institutional Arrangements, and Land-Use Planning," *Environment and Planning A*. 37/8: 1335-1352

Mitchell, B. (Ed.) 2004. *Resource and Environmental Management in Canada*. Don Mills: Oxford

Mitchell, B. and D. Draper. 1982. *Relevance and Ethics in Geography*. London: Longman

Mittelstaedt, M. 2001. "Ontario Dam at Risk of Collapse," *The Globe and Mail*. February 28

—2002. "Dead in the Water," *The Globe and Mail*. October 05

—2005. "Pembroke Factory Sparks Nuclear Concern" *The Globe and Mail*. November 03

—2007. "Critics Raise Red Flag over Fluoride in Drinking Water," *The Globe and Mail*. November 23

—2008. "Canadian Lakes Suffering from Aquatic Version of Osteoporosis," *The Globe and Mail*. November 28

—2009. "Green Crusaders Recycle Protest Tactics of Old," *The Globe and Mail*. July 13

Moench, M. 1999. "Addressing Constraints in Complex Systems—Meeting the Water Management Needs of South Asia in the 21st Century," in M. Moench et al. (Eds.) *Rethinking the Mosaic—Investigations into Local Water Management*. Kathmandu, Nepal: Water Conservation Foundation and Boulder, Colorado: Institute for Social and Environmental Transition

Moncton, City of. 2008. *Greater Moncton Water Action Plan*, <http://www.moncton.ca/Assets/Government%20English/Publications%20English/Greater%20Moncton%20Water%20Action%20Plan.pdf>

Montpetit, É. 2003. *Misplaced Distrust: Policy Networks and the Environment in France, the United States and Canada*. Vancouver, BC: UBC

Moraru-de Loë, L. and R. de Loë. 1998. "Water Management: Shifting Responsibilities," *Water News. Newsletter of the Canadian Water Resources Association*. 17/3: 1, 3

Morin, A. and B. Cantin. 2009. *Strengthening Integrated Water Resource Management in Canada*, 30 March 2012, <http://www.policyresearch.gc.ca/doclib/2009-0009-eng.pdf>

Moses, T. 1999. *"Water and First Nations,"* Canadian Bar Association National Symposium on Water Law, Toronto, April

Muldoon, P. and T. McClenaghan. 2007. "A Tangled Web: Reworking Canada's Water Laws," pp. 245-261 in Bakker, K. (Ed.) *Eau Canada: The Future of Canada's Water.* Vancouver: UBC

Muro, M. and P. Jeffrey. 2008. "A Critical Review of the Theory and Application of Social Learning in Participatory Natural Resource Management Processes," *Journal of Environmental Planning and Management.* 51/3: 325-344

Murray, J. 2004. *Water: Lawren Harris and the Group of Seven.* Toronto: McArthur

Murray, V. 2006. *The Nonprofit and Voluntary Sector in British Columbia. Regional Highlights from the National Survey of Nonprofit and Voluntary Organizations*, <http://www.imaginecanada.ca/files/en/nsnvo/g_british_columbia_sector_report.pdf>

Nanos, N. 2009. "Canadians Overwhelmingly Choose Water as our Most Important Natural Resource," *Policy Options.* July-August

National Round Table on the Environment and the Economy. 2010. *Changing Currents: Water Sustainability and the Future of Canada's Natural Resource Sectors*, <http://collectionscanada.gc.ca/webarchives2/20130322173842/http://nrtee-trnee.ca/water/water-sustainability-and-the-future-of-canadas-natural-resource-sectors-2/changing-currents>

—2011. *Paying the Price: The Economic Impacts of Climate Change for Canada*, <http://collectionscanada.gc.ca/webarchives2/20130322143132/http://nrtee-trnee.ca/wp-content/uploads/2011/09/paying-the-price.pdf>

—2012. *Moving to Action: NRT National Water Forum Report*, 15 <http://collectionscanada.gc.ca/webarchives2/20130323005245/http://nrtee-trnee.ca/wp-content/uploads/2013/02/NRT-Water-Forum-Report-eng.pdf>

Neizen, R. 1993. "Power and Dignity: The Social Consequences of Hydro-electric Development for the James Bay Cree," *Canadian Review of Sociology and Anthropology.* 30/4: 510-530

New Zealand. Environment. 2004. *Water Programme of Action. Water Allocation and Use*, <http://www.mfe.govt.nz/publications/water/water-allocation-use-jun04/water-allocation-use-jun04.pdf>

Niagara Peninsula Conservation Authority. 1999. *Welland River Watershed Strategy*, 17 May 2007, <http://www.conservation-niagara.on.ca/water_management/pdf/Welland%20River%20Strategy.pdf>

NiCHE (Network in Canadian History and Environment). 2007. *Confluences: A Workshop on Rivers, History and Memory*, <http://niche-canada.org/confluences>

Nieoczym, A. 2010. "Mudslide Prompts New Dam Safety Rules," *The Globe and Mail.* October 15

Nikiforuk, A. 2004. *Political Diversions: Annex 2001 and the Future of the Great Lakes*, <http://www.powi.ca/index_transboundary.php>

—2008. "Water, Tabled," *Globe and Mail Report on Business.* 24/9: 46-52

Norman, E. and K. Bakker. 2005. *Drivers and Barriers of Coopération in Transboundary Water Governance: A Case Study of Western Canada and the United States*, <http://www.watergovernance.ca/PDF/Gordon_Foundation_Transboundary_Report.pdf>

Norris, P. September 22, 2003. *Simplified Water Management Planning.* Letter to Members of the Ontario Waterpower Association, 23 August 2006, <http://www.owa.ca/OWASimple.pdf>

Northwest Territories Public Works and Services. 2005. *Managing Drinking Water Quality in the Northwest Territories*, <http://www.pws.gov.nt.ca/pdf/WaterAndSanitation/WaterFramework.pdf>

Norton, B. 2006. "Mark Sagoff's Price, Principle, and the Environment: Two Comments." *Ethics, Place and Environment*. 9/3: 337-372

Nowlan, L. 2004. *Customary Water Laws and Practice in Canada*, <http://www.fao.org/fileadmin/templates/legal/docs/CaseStudy_Canada.pdf>

—2005. *Buried Treasure: Groundwater Permitting and Pricing in Canada*, <http://www.nrcan.gc.ca/sites/www.nrcan.gc.ca.earth-sciences/files/pdf/gm/reports/pdf/gw_permitting_pricing_canada_e.pdf>

O'Connor, The Honourable D. 2002. *Part Two. Report of the Walkerton Inquiry: A Strategy for Safe Drinking Water*, <http://www.attorneygeneral.jus.gov.on.ca/english/about/pubs/walkerton/>

Okotoks, Town of. Planning Services. 2001. *Town of Okotoks Flood Plain Policy*, <http://www.okotoks.ca/data/1/rec_docs/184_FPP_2001_complete.pdf>

Oldman River Dam Environmental Advisory Committee. 2001. *Final Recommendations*, <http://environment.gov.ab.ca/info/library/6275.pdf>

Ontario Clean Water Agency. 2010. <http://www.ocwa.com/>

Ontario Environment. 2003. *Protecting Ontario's Drinking Water: Toward a Watershed-Based Source Protection Planning Framework, Final Report*, <http://www.ene.gov.on.ca/environment/en/resources/STDO1_076400.html>

—2005. *Permit to Take Water (PTTW) Manual*, <http://www.ene.gov.on.ca/environment/en/resources/STDO1_078778.html>

—2007. *Permit to Take Water (PTTW) Administrative Fees*, <http://www.ene.gov.on.ca/environment/en/resources/STDO1_078780.html>

Ontario Natural Resources. 2002. *Waterpower. Water Management Planning Guidelines for Waterpower*, <http://www.mnr.gov.on.ca/stdprodconsume/groups/lr/@mnr/@renewable/documents/document/251983.pdf>

Ouyahia, M. 2006. *Public-Private Partnerships for Funding Municipal Drinking Water Infrastructure: What Are the Challenges? Discussion Paper*, <http://www.pppcouncil.ca/pdf/muni_water_p3_report_govcanada_052006.pdf>

Ormsby, M. 1958. *British Columbia: A History*. Vancouver: Macmillan

Pacific Institute. 2006. *Water Data from the World's Water*, <http://www.worldwater.org/data.html>

Pahl-Wostl, C. 2002. "Towards Sustainability in the Water Sector—The Importance of Human Actors and Processes of Social Learning," *Aquatic Sciences*. 64: 394-411

Palacios, M. and J. Brown. 2005. "Saving Rivers through Water Transfers," *Fraser Review*, July, <http://www.fraserinstitute.org/research-news/research/display.aspx?id=11070>

Paperny, A. 2011. "Putting a Price on 'Acts of God'," *The Globe and Mail*. May 24.

Parr, J. 2004. "Local Water Diversely Known: Walkerton Ontario, 2000 and After," *Environment and Planning D: Society and Space*. 23/2: 251-271

Partners FOR the Saskatchewan River Basin. 2011. *About*, <http://www.saskriverbasin.ca>

Patrick, R. 2007. *Factors Facilitating and Constraining Source Water Protection in the Okanagan Valley, British Columbia: A Political Ecology Perspective.* Ph.D. Dissertation. Guelph, Ontario: University of Guelph

—2009. "Source Water Protection in a Landscape of 'New Era Deregulation,'" *Canadian Geographer.* 53/2: 208-221

Pearce, T. 2004. "Sip on This," *The Globe and Mail.* September 20

Pearse, P. 1998. "Water Management in Canada. The Continuing Search for the Federal Role," Victoria: Canadian Water Resources Association, Keynote Address at the 51st Annual Conference, Water News. *Newsletter of the Canadian Water Resources Association.* 17/3: 16-21

—2000. "Go with the Flow," *The Globe and Mail.* March 13

Pearse, P. and F. Quinn. 1996. "Recent Developments in Federal Water Policy: One Step Forward, Two Steps Back," *Canadian Water Resources Journal.* 21/4: 329-340

Penning-Rowsell, E. 1996. "Flood-Hazard Response in Argentina," *Geographical Review.* 86/1: 72-90

Percy, D. 1988. *The Framework of Water Rights Legislation in Canada.* Calgary: University of Calgary Canadian Institute of Resources Law

Perkel, C. 2002. *The Well of Lies: The Walkerton Water Tragedy.* Toronto: McClelland & Stewart

Perry, C. et al. 1997. *Water as an Economic Good: A Solution or a Problem,* <http://www.iwmi.cgiar.org/Publications/IWMI_Research_Reports/PDF/PUB014/REPORT14.PDF>

Peterson, C. 2006. "Canada Seeks Input on North Fork Coal," *Hungry Horse News,* December 14, <http://www.gravel.org/mt/archives/000208.html>

Peterson, T. 2003. "The Wet, Weird and Sometimes Wacky World of Water Ideas," *Aqueduct Magazine,* April, <http://www.mwdh2o.com/Aqueduct/april_2003/water_ideas.html>

Phillips, S. and M. Orsini. 2002. *Mapping the Links: Citizen Involvement in Policy Processes,* <http://www.eldis.org/go/country-profiles&id=11144&type=Document>

Pielou, E. 1998. *Fresh Water.* Chicago: University of Chicago

Pinch, P. and I. Munt. 2002. "Blue Belts: An Agenda for 'Waterspace' Planning in the UK," *Planning Practice & Research.* 17/2: 159-174

Pindera, G. 1997. "Red River Dance," *Canadian Geographic.* July/August: 52-62

Pinter, N. 2005. "One Step Forward, Two Steps Backward on U.S. Floodplains," *Science.* 308/5719: 207-208

Pires, M. 2004. "Watershed Protection for a World City: The Case of New York," *Land Use Policy.* 21: 161-175

Platt, R. 1997. "Classics in Human Geography Revisited," *Progress in Human Geography.* 21/2: 243-250

—1999. "From Flood Control to Flood Insurance: Changing Approaches to Floods in the United States," *Environments.* 27/1: 67-78

Pollon, C. 2010. "A River Lives through It," *The Globe and Mail.* August 21

Pollution Probe. 2007. *Towards a National Vision and Strategy for Water Policy in Canada,* <http://www.

waterbucket.ca/wcp/sites/wbcwcp/documents/media/55.pdf>

Pope, A. "Québec Planning Flood Clean-Up," *The Weather Network*, <http://www.theweathernetwork.com/news/storm_watch_stories3&stormfile=quebec_planning_flood_cleanu_290511?ref=ccbox_weather_category1>

Porteous, J. 1996. *Environmental Aesthetics. Ideas, Politics and Planning*. London: Routledge

Porteous, J. and S. Smith. 2001. *Domicide: The Global Destruction of Home*. Montreal: McGill-Queen's

Porter, M. c1999. *Use of Groundwater Resources on Mayne Island, BC*. Victoria, BC: University of Victoria Unpublished Master's thesis

Pratt, J. 1999. *Water Law in Newfoundland and Labrador*. Canadian Bar Association National Symposium on Water Law. Toronto, April

Priest, L. 2005. "Reserve's Medical Emergency Did Not Come as a Surprise," *The Globe and Mail*. October 29

Prince Edward Island. Environment and Energy. 2003. *Clear from the Ground to the Glass*, <http://www.gov.pe.ca/photos/original/10points2purity.pdf>

Prince Edward Island. Fisheries, Aquaculture and Environment. c2001. *Home Heat Tank Safety*, <http://www.gov.pe.ca/photos/original/HomeHeatSafety.pdf>

Prowse, T. et al. 2003. *Dams, Reservoirs and Flow Regulation*, <http://www.ec.gc.ca/INRE-NWRI/default.asp?lang=En&n=0CD66675-1&offset=7&toc=show>

Purdy, A. 2000. "Say the Names," *Beyond Remembering*. Madeira Park, BC: Harbour

Québec. 2006. *Using Energy to Build the Québec of Tomorrow*, <http://www.mrnf.gouv.qc.ca/english/publications/energy/strategy/energy-strategy-2006-2015-summary.pdf>

Quinn, D. 2007. "Flathead on the Mind," *Westworld*. Summer: 40-45

Quinn, F. 1991. "As Long as the Rivers Run: The Impacts of Corporate Water Development on Native Communities in Canada," *The Canadian Journal of Native Studies*. XI/1: 137-154

—2007. *Water Diversion, Export and Canada-US Relations: A Brief History*. Toronto: Munk Centre for International Studies Program on Water Issues, <http://powi.ca/wp-content/uploads/2012/12/Water-Diversion-Export-and-Canada-US-Relations-A-Brief-History-2007.pdf>

Quinn, F. et al. 2003. *Water Allocations, Diversion and Export*, <http://www.ec.gc.ca/inre-nwri/default.asp?lang=En&n=0CD66675-1&xml=0CD66675-AD25-4B23-892C-5396F7876F65&offset=6&toc=show>

Ramin, V. 2004. "The Status of Integrated Water Resources Management in Canada," *Canadian Perspectives on Integrated Resources Management*. Cambridge, Ont.: Canadian Water Resources Association

Reiser, D. et al. 1989. "Status of Instream Flow Legislation and Practices in North America," *Fisheries*. 14: 22-29

Regina Leader Post. 2002. "Water Issues Have Priority," *Regina Leader Post*, 30 April

Reguly, E. 2006. "Water Exports? Harper Should Keep the Taps Closed," *The Globe and Mail*. January 21

Reid, S. 2004. "Children's Groundwater Festivals," *Ribbons of Life: Celebrating the Past—Charting the Future*. 4th Canadian River Heritage Conference, <http://www.grandriver.ca/RiverConferenceProceedings/ReidSusan.pdf>

Renzetti, S. 2005. "Economic Instruments and Canadian Industrial Water Use," *Canadian Water Resources Journal*. 30/1: 21-30

Renzetti, S. and C. Busby. 2009. "Water Pricing: Infrastructure Grants Hinder Necessary Reform," *Policy Options*. 30/7: 32-35

RésEau. 2010. *RésEau*. <http://www.reseauwaternet.ca/>

Richardson, B. 2008. *Strangers Devour the Land*. Revised Edition. White River Junction, VT: Chelsea Green

Rivera, A. 2005. "How Well Do We Understand Groundwater in Canada? A Science Case Study," pp. 4-12 in Nowlan, L. 2005. *Buried Treasure: Groundwater Permitting and Pricing in Canada*, <http://www.nrcan.gc.ca/sites/www.nrcan.gc.ca.earth-sciences/files/pdf/gm-ces/reports/pdf/gw_permitting_pricing_canada_e.pdf>

Rivera, A. et al. 2003. *Canadian Framework for Collaboration on Groundwater*. Ottawa: Government of Canada

Roach, R. et al. 2004. *Drop by Drop: Urban Conservation Practices in Western Canada*. Western Cities Project Report #29, <http://cwf.ca/publications-1/drop-by-drop-urban-water-conservation-practices-in-western-canada>

Rogers, E. et al. 1997. "A Paradigmatic History of Agenda-Setting Research," pp. 225-236 in S. Iyengar and R. Reeves (Eds.) *Do The Media Govern?* Thousand Oaks, California: Sage

Rogers, P. et al. 2002. "Water is an Economic Good: How to Use Prices to Promote Equity, Efficiency, and Sustainability," *Water Policy*. 4: 1-17

Rosen, C. and J. Tarr. 1994. "The Importance of an Environmental Perspective in Urban History," *Journal of Urban History*. 20/3: 299-310

Roth, W-M. et al. 2004. "Those Who Get Hurt Aren't Always Being Heard: Scientist-Resident Interaction over Community Water," *Science, Technology & Human Values*. 29/2: 153-218

Roué, M. 2003. "US Environmental NGOs and the Cree. An Unnatural Alliance for the Preservation of Nature?" *International Social Science Journal*. 55/178: 619-627

Sagoff, M. 1974. "On Preserving the Natural Environment," *Yale Law Journal*. 84/2: 205-267

Salinas, E. 2006. "Tofino Wedding Plans Evaporate in Drought," *The Globe and Mail*. August 31

Sandercock, L. 1998. *Towards Cosmopolis*. New York: Wiley

Sarton, M. 1968. *Plant Dreaming Deep*. New York: Norton

Sandink, D. et al. 2010. *Making Flood Insurable for Canadian Homeowners*, <http://www.iclr.org/flood-droughtinsurable.html>

Saskatchewan Watershed Authority. 2003. *Protecting Our Water. A Watershed and Aquifer Planning Model for Saskatchewan*, <http://www.publications.gov.sk.ca/deplist.cfm?c=998&d=126&cl=2>

Saskatchewan Watershed Authority and the Lower Souris River Watershed Committee. 2006. *Lower*

Souris River Watershed. Source Water Protection Plan, <https://www.wsask.ca/Global/Water%20Info/Watershed%20Planning/LowerSourisRiverWatershedSourceWaterProtectionPlan.pdf>

Saunders, J. 1988. *Interjurisdictional Issues in Canadian Water Management*. Calgary: Canadian Institute of Resources Law

Saunders, J. and M. Wenig. 2007. "Whose Water?" pp. 119-141 in Bakker, K. (Ed.) *Eau Canada. The Future of Canada's Water.* Vancouver: UBC

Savenije, H., 2002. "Why Water is not an Ordinary Economic Good, or Why the Girl is Special", *Physics and Chemistry of the Earth*, 27/11-22: 741-744

Savoie, D. 2010. *Power: Where is It?* Montréal: McGill-Queen's

Schindler, D. 2006. *The Myth of Abundant Canadian Water*, <http://waterintheworks.wordpress.com/2009/12/03/the-myth-of-abundant-canadian-water-david-schindler/>

—2010. "Keynote Address," *Canadian Water: Towards a New Strategy*. McGill Institute for the Study of Canada Conference. March 24, <http://www.mcgill.ca/water2010/>

Schindler, D. et al. 2008. "Eutrophication of Lakes Cannot Be Controlled by Reducing Nitrogen Input: Results of a 37-Year Whole-Ecosystem Experiment," *Proceedings of the National Academy of Sciences*. 105/ 32: 11254-11258

Schornack, D. and J. Nevin. 2006. *The International Joint Commission: A Case Study in the Management of International Waters*. Rosenberg International Forum on Water Policy, <http://rosenberg.ucanr.org/forum5.cfm?displaysection=4>

Scoffield, H. 1998. "Delay of Export Law Called a Threat to Water Resources," *The Globe and Mail*. November 02

Scoffield, H. and D. LeBlanc. 1999. "Ban Near on Export of Fresh Water," *The Globe and Mail*. October 02

Scott, A. 1991. "British Columbia's Water Rights: Their Impact on the Sustainable Development of the Fraser Basin," pp. 341-390 in A. Dorcey (Ed.) 1991. *Perspectives on Sustainable Development in Water Management: Towards Agreement in the Fraser River Basin*. Vancouver: UBC Westwater Research Centre

Scruton, D. et al. 1998. "Pamehac Brook: A Case Study of the Restoration of a Newfoundland, Canada River Impacted by Flow Diversion for Pulpwood Transportation," *Aquatic Conservation: Marine and Freshwater Ecosystems*. 8/1:145-47

Séguin, R. 2006. "Troubled Waters," *The Globe and Mail*. December 19

—2011. "Innu and Hydro-Québec Reach Deal on $6.5 Billion Project," *The Globe and Mail*. January 25

Sewell, W. 1969. "Human Response to Floods," pp. 431-451 in R. Chorley and R. Barry (Eds.). *Water, Earth and Man*. London: Methuen

Shiva, V. 2002. *Water Wars. Privatization, Pollution and Profit*. Toronto: Between the Lines

Shordt, K. 2000. "Monitoring in a Historic Perspective," in Water Supply and Sanitation Council. *Vision 21—Water for People*, 16 February 2000, <http://ww.wsscc.org/vision21/docs/doc29.html>

Shoumatoff, A. 2005. "Who Owns This River?" *One Earth*, <http://www.nrdc.org/onearth/05spr/manitoba1.asp#>

Shrubsole, D. 2001. *Virtually Untapped: Water Demand Management in Ontario.* London, Ontario: University of Western Ontario

Shrubsole, D. (Ed.) 2004. *Canadian Perspectives on Integrated Resources Management.* Cambridge, Ont.: Canadian Water Resources Association

Shrubsole, D. and B. Mitchell (Eds.) 1997. *Practicing Sustainable Water Management: Canadian and International Experiences.* Cambridge, Ont.: Canadian Water Resources Association

Shrubsole, D. et al. 2000. *Flood Management in Canada. At the Crossroads.* Toronto: Institute for Catastrophic Loss Reduction, <http://www.iclr.org/images/Flood_Management_in_Canada_at_the_Crossroads.pdf>

—2003. *An Assessment of Flood Risk Management in Canada,* <http://wsm.ezsitedesigner.com/share/scrapbook/42/425698/An_Assessment_of_Flood_Risk_Management_in_Canada.pdf>

Sierra Legal (now Ecojustice). 29 November 2006. "US, Canadian Cities Fouling the Great Lakes with Raw Sewage," *Media Release,* <http://www.ecojustice.ca/media-centre/press-releases/us-canadian-cities-fouling-the-great-lakes-with-raw-sewage>

Skaburskis, A. 1988. "Criteria for Compensating the Impacts of Large Projects: The Impact of British Columbia's Revelstoke Dam on Local Government Services," *Journal of Policy Analysis and Management.* 7/4: 668-686

Slocombe, D. 2004. "Applying an Ecosystem Approach," pp. 442-466 in Mitchell, B. (Ed.) 2004. *Resource and Environmental Management in Canada.* Don Mills, Ontario: Oxford

Small Hydro International Gateway. Accessed 2012. *Canada. Country Brief,* <http://www.small-hydro.com/Past-Contributors-Pages/Canada.aspx>

Smith, D. 1999. "Conclusion: Towards a Context-Sensitive Ethics," pp. 275-290 in Procter, J. and D. Smith (Eds.) *Geography and Ethics: Journeys in a Moral Terrain.* London: Routledge.

Smith, S. 1991. "Floodplain Management in the Fraser Basin," pp. 115-132 in Dorcey, A. (Ed.) 1991. *Perspectives on Sustainable Development in Water Management: Towards Agreement in the Fraser Basin.* Vancouver: UBC Westwater Research Centre

—1995. *Domicide: Concept, Experience, Planning.* Victoria, BC: University of Victoria Unpublished PhD Dissertation

—2001. "Water Resources," pp. 65-81 in C. Wood, (Ed.) *British Columbia, The Pacific Province: Geographical Essays.* Victoria: University of Victoria Department of Geography Western Geographical

Sneddon, C. et al. 2002. "Contested Waters: Conflict, Scale, and Sustainability in Aquatic Socio-ecological Systems," *Society & Natural Resources.* 15/8: 663-675

Sproule-Jones, M. 2002 "Institutional Experiments in the Restoration of the North American Great Lakes," *Canadian Journal of Political Science.* 35/4: 835-857

—2003. *Restoration of the Great Lakes: Promises, Practices and Performances.* Vancouver: UBC

—2008. "Property Rights and Water," pp. 116-130 in Sproule-Jones, M. et al (Eds.) *Canadian Water Politics: Conflicts and Institutions.* Montréal: McGill-Queen's

Sproule-Jones, M. et al (Eds.) 2008. *Canadian Water Politics: Conflicts and Institutions.* Montréal: McGill-Queen's.

Spulber, N. and A. Sabbaghi. 1994. *Economics of Water Resources: From Regulation to Privatization.* Boston: Kluwer Academic

St. Lawrence Vision 2000. 2003. "St. Lawrence Vision 2000 Five-Year Report 1998-2003: Tangible Environmental Achievements Attained Thanks to a Partnership and Consultation-based Approach," *Press Release*, <http://www.mddefp.gouv.qc.ca/communiques_en/c20030826-slv2000.htm>

Stenson, F. May/June 2003. "Cool, Clear Water," *Alberta Views*, <http://albertaviews.ab.ca/issues/2003/mayjun03/mayjun03wit.pdf>

Stott, R. and L. Smith. 2001. *River Recovery: Restoring Rivers and Streams through Dam Decommissioning and River Modification.* B.C. Outdoor Recreation Council, <http://www.recovery.bcit.ca/pdfs/final_report.pdf>

Strauss, J. 2005a. "A Hero's Welcome for Kashechewan Chief," *The Globe and Mail.* October 29

—2005b. "To Stem the Tide," *The Globe and Mail.* September 15

Stream of Dreams Mural Society. 2011. *Stream of Dreams. Watershed Education Through Community Art,* <http://www.streamofdreams.org/>

Sustainable Prosperity. 2012. *Environmental Markets 2012,* <http://www.sustainableprosperity.ca/article3228>

Swain, H. 2009. "Drinking Water and Waste Water: A Primer," *Policy Options.* 30/7: 26-30

Swain, H. et al. 2006. *Report of the Expert Panel on Safe Drinking Water for First Nations,* <http://www.safewater.org/PDFS/reportlibrary/P3._EP_-_2006_-_V1.pdf>

Swiss Re. 2003. *Focus Report: Dams,* 15 January 2011, <http://www.swissre.com/>

Tabara, D. et al. 2005. *Sustainability Learning for River Basin Management and Planning in Europe,* <http://www.harmonicop.uni-osnabrueck.de/_files/_down/WP6%20Integration%20reportFINAL.pdf>

Tate, D. 2000. *Principles of Water Use Efficiency,* <http://www.bvsde.paho.org/muwww/fulltext/repind48/principles/principles.html>

Taylor, A. 2003. *Drinking Water Protection Law in British Columbia.* Canadian Bar Association National Symposium on Water Law. Vancouver, BC: Canadian Bar Association, March 28, 29

Terry, J. and A. Unikowsky. 2003. *Application of the NAFTA and the GATT to Water.* Canadian Bar Association National Symposium on Water Law. Vancouver, BC: Canadian Bar Association, March 28, 29

TheRecord.com. 2009. "Conestogo Dam Improvements Announced," *TheRecord.com*, 29 June. 04 November 2010, <http://news.therecord.com/article/561243>

Theodore, T. 2003. "B.C. Government Subject of Damning Report Regarding Drinking Water Safety," *Canadian Press,* October 09

Thiele, L. 2000. "Limiting Risks: Environmental Ethics as a Policy Primer," *Policy Studies Journal.* 28/3: 540-557

Theriault, R. 2006. "Assessing the Water Quality Benefits of BMPs at Watershed Scale Across Canada," *Working from the Source towards Sustainable Water Management.* Canadian Water Resources Association Conference, 10 July 2006, <http://www.allsetinc.com/cwra2/files/presentations/3B_George_Theriault.pdf>, see <http://www4.agr.gc.ca/AAFC-AAC/display-afficher.do?id=1181580137261&lang=eng>

Thomas, A. 2004. *Framing the Debate: Media Coverage of the Water Export Issue*. Victoria, B.C.: University of Victoria Department of Geography Directed Studies

Thomas, M. 2013. "Future Floods in Calgary," *Urban WorkBench. Sustainable Designs for Life*, <http://urbanworkbench.com/>

Thompson, S. 1999. *Water Use, Management and Planning in the United States*. San Diego: Academic

Thoreau, H. in R. Sattlemayer (1987) "Reflections on a Child's Water Wheel," *The Missouri Review*. 10/2

Tonn, B. et al. 2000. "A Framework for Understanding and Improving Environmental Decision Making," *Journal of Environmental Planning and Management*. 43/2: 163—183

Trumbull, M. 2005. "Cost of Katrina to Hit Budget Hard," *Christian Science Monitor*, September 14, <http://www.csmonitor.com/2005/0914/p01s01-usec.html>

Tuan, Y-F. 1999. "Geography and Evil: A Sketch," pp. 106-119 in Procter, J. and D. Smith (Eds.) *Geography and Ethics: Journeys in a Moral Terrain*. London: Routledge.

Tully, S. 2000. "Water, Water Everywhere," *Fortune*. 141/10: 342-354

Tutton, M. 2007. "Investigators Blame Pesticide Runoff for Massive Fish Kill," *The Globe and Mail*. July 24

Tyler, K. 1999. *The Division of Powers and Aboriginal Water Rights Issues*. Canadian Bar Association National Symposium on Water Law. Toronto: April

United Nations Development Program. Young People of the World. 2006. *Water Rights and Wrongs*, <http://hdr.undp.org/external/hdr2006/water/index.htm>

UN Water. 2009. *The 3rd United Nations World Water Development Report: Water in a Changing World*, <http://www.unesco.org/new/en/natural-sciences/environment/water/wwap/wwdr/wwdr3-2009/>

US Department of the Interior. Bureau of Reclamation. 2006. *Comment Letters, Draft Report on the Red River Valley*, <http://www.usbr.gov/gp/dkao/redriver/rrvwsp/Appendixes/letters/index.cfm>

US Environmental Protection Agency. 2003. *Principles of Watershed Management*, <http://www.epa.gov/OWOW/watershed/wacademy/acad2000/watershedmgt/index.html>

US Federal Emergency Management Agency. 2004. *The National Flood Insurance Program*, <http://www.fema.gov/about/programs/nfip/index.shtm>

US Geological Survey. 2012. *Devils Lake Basin in North Dakota*. <http://nd.water.usgs.gov/devilslake/>

University of Alberta. Environmental Research and Studies Centre and University of Toronto. Munk Centre for International Studies. Program on Water Issues. 2007. *Running out of Steam? Oil Sands Development and Water Use in the Athabasca River-Watershed: Science and Market based Solutions*, <http://powi.ca/wp-content/uploads/2007/05/Final-Running-out-of-Steam-Meeting-Notes.pdf>

University of British Columbia. Faculty of Forestry. 2002. *Results Based Forest and Range Practices for British Columbia. 6.0 Results-based Management Regimes for Specific Forest and Environmental Values*, <http://www.for.gov.bc.ca/code/discussionpaper/06.htm>

University of Victoria. Environmental Law Clinic. 2010. *Reinventing Rainwater Management: A Strategy to Protect Health and Restore Nature in the Capital Region,* <http://www.elc.uvic.ca/press/documents/stormwater-report-FINAL.pdf>

University of Victoria. Facilities Management. 2004. *Sustainability Report 2003~2004,* 30 April 2007, <http://web.uvic.ca/fmgt/assets/pdfs/FACI%20Sustain%20report.pdf>

Updike, J. 1989. *Just Looking: Essays on Art.* New York: Knopf

Upper Thames Valley Conservation Authority. 2007. *Flooding on the Thames River,* <http://www.thamesriver.on.ca/Water_Management/flood_history.htm>

Van Veen, D. et al. 2003. "Selecting Appropriate Dispute Resolution Techniques: A Rural Water Management Example," *Applied Geography.* 23/2-3: 89-113

Vidal, J. 2006. "Running on Empty," *The Guardian Weekly.* September 29

Vyvyan, C. 1998. *The Ladies, the Gwich'in and the Rat. Travels on the Athabasca, Mackenzie, Rat, Porcupine and Yukon Rivers in 1926.* Edmonton: University of Alberta

Waddell, I. 2002. *A Thirst to Die For.* Edmonton: Newest

Walkem, A. 2007. "The Land is Dry: Indigenous Peoples, Water and Environmental Justice," pp. 303-319 in Bakker, K. (Ed.) 2007. *Eau Canada: The Future of Canada's Water.* Vancouver: UBC

Walton, D. 2001. "Crusader for Alberta Dam Presses On," *The Globe and Mail.* August 20

Waltner-Toews, D. 2005. *Links between Climate, Water and Waterborne Illness, and Projected Impacts of Climate Change,* <http://hc-sc.gc.ca/ewh-semt/pubs/climat/newsletter-bulletin-2/research-recherche_2-eng.php>

Walz, L. 2009. "Theodosia Roundtable Takes Off," *PeakOnline,* <http://www.prpeak.com/articles/2009/12/04/news/doc4b15cff8ebc98619491512.txt>

Warwick, S. (2013) "Watermarks," *The Globe and Mail.* June 29.

Water Canada. 2010. "Conoco Phillips to Donate Medicine River Water Licence," *Water Canada,* June 09, <http://watercanada.net/2010/conocophillips-to-donate-medicine-river-water-licence/>

Water Policy and Governance Group. 2012. *Understanding the Current and Potential Role of the Natural Resources Sector in Collaborative Approaches to Water Governance: Summary of Survey Findings,* <http://www.wpgg.ca/news/211>

waterbucket.ca. 2006. *Water-Centric Planning Community-Of-Interest,* <http:/www.waterbucket.ca>

Webb, C. 2004. *Performance Measurement of Water Conservation Programs by British Columbia Water Utilities: A Report.* University of Victoria Department of Geography Directed Studies Report

Wellington Water Watchers. 2007. *Submission on Nestlé Aberfoyle Application,* <http://www.wellingtonwaterwatchers.ca/wp-content/uploads/2011/02/letter-to-MOE-revisions-Feb-16-rev1-1.pdf>

Wiewel, W. "Comment 4," in J. Friedman et al. 2004. "Strategic Planning and the Longer Range," *Planning Theory and Practice.* 5/1: 49-67

Westcoat, J. 1992. "Beyond the River Basin: The Changing Geography of International Water Problems and International Watercourse Law," *Colorado Journal of International Environmental Law & Policy.* 3: 310-330

Western Canada Water. 2012. *Strategic Plan. Working Together for Water,* <http://wcwwa.ca/documents/2012/08/wcw-strategic-plan-handout.pdf>

Westwater Research Centre. 1972. "Man, Floods, and the Environment," *Westwater Notes on Water Research in Western Canada.* No.2. University of BC: Photocopy

White, G. 1945. *Human Adjustment to Floods.* University of Chicago Department of Geography Research Paper No. 29. Chicago: University of Chicago Department of Geography

—1994. "Testimony and Prepared Statement," *Midwest Floods of 1993: Flood Control and Floodplain Policy and Proposals: Hearing before the Subcommittee on Water Resources and Environment, October 27, 1993,* <http://www.colorado.edu/hazards/gfw/quotes.html>

Whitten, R. 2006. *As Long As the Rivers Flow: Rural & First Nations Community Collaboration in Watershed Stewardship.* Canadian Water Resources Association Conference, 25 September 2006, <http://www.allsetinc.com/cwra2/files/presentations/3D_Reg_Whitten.pdf>

Windsor, J. (Ed.) 1992. *Water Export: Should Canada's Water Be for Sale?* Cambridge, Ont.: Canadian Water Resources Association

—1993. "Water Diversions and Export: Threat or Opportunity," pp. 227-277 in Foster, H. (Ed.) 1993. *Advances in Resource Management: Tributes to W.R. Derrick Sewell.* London: Belhaven

Windsor, J. and J. McVey. 2005. "Annihilation of Both Place and Sense of Place: The Experience of the Cheslatta T'En Canadian First Nation within the Context of Large-Scale Environmental Projects," *The Geographical Journal.* 171/2: 146-165

Wismer, S. and B. Mitchell. 2005. "Community-Based Approaches to Resource and Environmental Management," *Environments Journal.* 33/1: Theme Issue Introduction.

Wolf, A. 1998. "Conflict and Cooperation along International Waterways," *Water Policy.* 1/2: 251-265

Won, S. 2007. "Funds Turn to Infrastructure Investments," *The Globe and Mail.* July 09

Wood, D. 2002. "Latest Plan to Ease Water Woes: Big Baggies," *Christian Science Monitor.* 94/74: 1

Working Group on Domestic Reclaimed Water of the Federal-Provincial-Territorial Committee on Health and the Environment. 2010. *Canadian Guidelines for Domestic Reclaimed Water for Use in Toilet and Urinal Flushing,* <http://www.hc-sc.gc.ca/ewh-semt/pubs/water-eau/reclaimed_water-eaux_recyclees/index-eng.php>

World Commission on Dams. 2000. *Dams and Development: A New Framework for Decision Making. Report of the World Commission on Dams,* <http://www.internationalrivers.org/files/attached-files/world_commission_on_dams_final_report.pdf>

World Health Organization. 2001. World Water Day 2001. *Water, Health and Human Rights,* <http://www.who.int/water_sanitation_health/en/humanrights.html>

Wouters, P. et al. 2000. *The Legal Response to the World's Water Crisis: What Legacy from The Hague? What Future in Kyoto?* 30 March 2012, <http://www.africanwater.org/Documents/colorado_draft_4.doc>

Wright, A. 2003. "After Taking a Pounding, Hydro Is Ready for Comeback," *ENR: Engineering News-Record.* 250/25: June 30

Yakabuski, K. 2005. "Big Dreams in Canada's City that Never Sleeps," *The Globe and Mail.* July 16

Zbarsky, M. 2005. *Implementation of Combined Land Use and Water Resource Management Plans in British Columbia: Factors for Success and the Role of the Community.* Victoria, BC: University of Victoria Department of Geography Honours Paper

Zellner, M. 2008. "Embracing Complexity and Uncertainty: The Potential of Agent-Based Modeling for Environmental Planning and Policy," *Planning Theory and Practice.* 9/4: 437-457.

APPENDIX 1: GOVERNMENTS' WATER MANAGEMENT ROLES

Federal Government

The federal role began even before Confederation with interest in ports and shipping. This expanded given the trans-jurisdictional nature of surface water and groundwater, government's ability to coordinate interprovincial matters, the need to protect national and international aquatic and terrestrial species, and the importance of water as a "national concern"(Saunders and Wenig 2007, 120-121). Provincial and territorial governments share formal powers under the *Constitution Act 1867, s.91*, although there is no mention of the word "water".[122] The federal government's general provisions relating to "spending power", and "peace, order and good government" are augmented by legislative jurisdiction over fisheries (91.12), navigation and shipping (91.10), and criminal law (91.29), as well as by powers affecting aboriginal peoples, taxation, trade and international matters, agriculture (shared with the provinces) and aspects of water quality (Saunders 1988, 10; 2007, 122).

Federal-provincial agreements to carry out these responsibilities include cost-sharing, water surveys, flood damage reduction, and Water Boards (for example, the Prairie Provinces Water Board and MacKenzie River Board). The *International River Improvements Act* (1985) governs international flows, giving authority to licence activities that may alter the flow of rivers into the United States. The *Boundary Waters Treaty of 1909* establishes the International Joint Commission (IJC). Of particular importance has been the IJC's work relating to water quality of the Great Lakes and the issue of bulk water removal (International Boundary Waters Treaty Act 1985).

The federal role was defined by various initiatives: the *Canada Water Act* (1970)[123], the *Arctic Waters Pollution Prevention Act*, and the *Northern Inland Waters Act*. The *Great Lakes Water Quality Agreement* was signed in 1972. Under this agreement between the United

States and Canada, efforts first targeted point source pollution, and then began in 1985 to identify "areas of concern" and prepare remedial action plans for pollution reduction. In 1975, the National Flood Damage Reduction Program was approved. Federal programs have supported the St. Lawrence Seaway and Power Project, the South Saskatchewan Project, the Salmonid Enhancement Program, the Fraser River Flood Control Program, and the Floodplain Mapping Agreement, as well as planning studies.

The *Department of the Environment Act* (1985) makes the Minister of Environment the national leader for water management. To further define the federal role, the Inquiry on Federal Water Policy of 1984-85 resulted in the 1987 Federal Water Policy with a focus on water pricing, science leadership, integrated planning, legislation, and public awareness (Canada Environment 1987). The *Canadian Environmental Assessment Act* (1995) enables review of projects that could have serious environmental effects on another province. The *Canadian Environmental Protection Act* (1999) strengthens the federal government's role regarding pollution prevention and elimination of the most persistent and bio-accumulative toxic substances. This Act also recognizes the importance of an ecosystem approach and enables environmental standards, ecosystem objectives and environmental quality guidelines and codes of practice.

Many have called for a greater federal role and enhanced leadership yet, in general, the federal interest has narrowed. Reacting to budget cuts in the late 1980s and through the 1990s, the federal government argued that natural resources are a provincial responsibility (Bruce 2006, 2). Federal interests then focused on sustainable development and aquatic resource management, promotion of water-efficient technologies, national standards for water quality and environmental monitoring, and leadership in fresh water science and data collection. Attention was also given to environmental issues at the inter-provincial level, such as management of the Mackenzie River, and, at the international level, the IJC and climate change (Pearse 1998). More recent efforts have addressed the issue of water export through the IJC (see Chapter 8).

As governments change, so do priorities. Environment Canada now delivers a much depleted "water" focus, sharing responsibility with nineteen federal departments (Côté 2004, 6). Environment Canada's *Sustainable Development Strategy* (2010) sets specific targets relating to water quantity and quality. Another significant focus is climate change and adaptation. The National Water Research Institute (National Hydrology Research Centre, the Canadian Centre for Inland Waters and the Water and Climate Impacts Research Centre) is the main organization identifying threats to Canada's water quality and quantity, including the potential impact of climate change. Sustainable Development Technology Canada funds innovative technology relating to projects addressing clean water (and other sustainability parameters). The Internet project RésEau makes water information more accessible (RésEau 2010). Certain areas are the subject of Action Plans: the Georgia Basin, the St. Lawrence, the Northern Ecosystem, the Western Boreal Conservation Initiative, and the Great Lakes. The Action Plan on Clean Water targets clean-up of waterbodies and drinking water provision for aboriginal communities.

The 1972 Great Lakes Water Quality Agreement (1972, 1978) and the 1987 Protocol (Remedial Action Plans for specific areas and clean-up of contaminated sediments) display conflicting results (Botts and Muldoon 2005, 187). The Agreement reduced phosphorus and other pollutants and toxic contaminants, promoted an ecosystem approach, augmented science in these areas, and fostered a Great Lakes community and partnership with the U.S. (Botts and Muldoon 2005, 188). But commentators suggest that Canada's activity to date is "meagre" (Muldoon and McClenaghan 2007, 249). A 2012 amendment to the Agreement will hopefully go some way to rectifying this situation (Canada Environment 2012).

Provincial and Territorial Governments

There are overlaps in jurisdiction between the federal and provincial/territorial governments; for example, where transboundary flows affect federal lands or the federal fisheries mandate overlaps with provincial/territorial rights. These matters may be resolved by intergovernmental agreement, consistent with the philosophy of Canadian federalism that provincial autonomy should be respected (Saunders 1988, 36-37; 2007, 121). But effectiveness of intergovernmental institutions varies from region to region and depending on the type of conflict (Johns 2009, 80).

A major direction for water management has come from provincial governments as they exercise their powers under the Constitution Act 1867, s.92 (legislative powers that can be used for the purposes of water management), and s.109 (proprietary rights vesting ownership of lands, mines, minerals and royalties) (Saunders 1988, 9). Territorial governance is discussed in more detail in chapters on water allocation and water quality. Provincial government activity in water management is not surprising given their level of autonomy, readiness to assume responsibility, and the absence of federal construction agencies such as the United States Army Corps of Engineers.[124] Then there are the very practical aspects: provinces have better access to the problems at hand and means to resolve them (Saunders and Wenig 2007, 121).

Provincial governments' activities apply to both surface and groundwater and include data collection, allocating water among competing users, promoting efficient water use, regulating the protection of water quality and use of thermal and hydroelectric development, promoting flood damage reduction and watershed planning, and enforcement of laws and regulations. The provinces consider water in the development of other resources and assess environmental impact and other consequences. Some programs are devolved to other agencies. At the provincial level, and sometimes at the regional level, health agencies play an increasing role in drinking water protection. Echoing federal government experience, provincial budgets are cut and activities within mandates change. A very active period in the 1980s to mid-1990 delivered services and advice. Now there is more emphasis on developing environmental standards and performance expectations, monitoring and public reporting, and ensuring positive compliance.

Officers of Parliament

Also notable at the federal and provincial levels is the emergence of Officers of Parliament such as the Auditor-General; on behalf of the Auditor General, the Commissioner of the Environment and Sustainable Development; the Privacy Commissioner and the Ombudsman. Although at arms-length from the decision-making process, their influence is considerable. Similarly, special inquiries such as the Honourable Justice O'Connor's Walkerton Inquiry change water governance and practice.

Interprovincial Bodies

The Canadian Council Ministers of Environment (CCME), the Council of the Federation, two water boards, and a number of committees exemplify interprovincial/territorial cooperation on shared water resources supported by the federal government. The CCME determines national environmental priorities and develops useful cross-country approaches. The Council of the Federation was formed by Canada's Premiers to foster collaborative intergovernmental relations. Its Water Stewardship Council provides information and strategic advice to the Premiers and promotes the Federation Water Charter.

The Prairie Provinces Water Board, formed in 1948, now apportions interprovincial waters, acts as a centre for information exchange, and generally promotes cooperation. The Mackenzie River Basin Board was formed by the Governments of Canada, British Columbia, Alberta, Saskatchewan, the Northwest Territories and the Yukon. The Master Agreement of 1997 has not been finalized with an operational agreement on transboundary watercourses' quality and quantity as yet (Gordon Foundation 2011). Programs such as the National Water Supply Expansion Program provide funds to help the agricultural sector prepare for water shortages.

Special Authorities

Examples of special authorities are the Halifax Regional Water Commission, responsible for water supply (discussed in Chapter 7) and Ontario's conservation authorities. Conservation authorities are stand-alone agencies that began with an emphasis on flooding and erosion concerns and now are 36 "community-based environmental organizations dedicated to conserving, restoring, developing and managing natural resources on a watershed basis" (Conservation Ontario 2011).

Aboriginal Water Governance

Governance of water for aboriginal peoples depends on law and agreements governing aboriginal and treaty rights, including the results of land claims. For the Yukon, Northwest and Nunavut Territories, the federal government carries out programs under the purview of Indian and Northern Affairs Canada. Their primary responsibility is to support First Nations and Inuit in joint management to develop healthy, sustainable communities and achieve economic and social goals. Water Boards carry out relevant responsibilities, the autonomy of each is dependent on the land claims agreement and implementing legislation.

Local Water Governance

Many local governments in Canada supply fire protection, drinking water, and waste treatment systems; Ball (1988) traces the history of public works, reminding us that before 1850, most of urban Canada existed without running water or indoor toilets and with flooded streets and basements. Saint John, Montreal, Toronto and Halifax were exceptions. In 1854 a cholera epidemic in Hamilton, Ontario killed nearly 4 percent of the town's residents and spurred the city into action; a waterworks and pumping station was built. As protectors of the environment, local governments continue to expand their roles. While the creation of parks and leave strips along watercourses and around lakes is common, local governments now enact many other protection regulations and plans. Local community groups augment their role. Together, local governments participate in the Federation of Canadian Municipalities. This body represents the interests of municipalities on policy and program matters that fall within federal jurisdiction and assists by conducting policy research (for example, on the use of tap water rather than bottled water at municipal facilities).

ENDNOTES

(Unless otherwise noted, website addresses are as of May 31, 2013.)

Chapter 1:

1   2012: The Forum for Leadership on Water's cross-Canada forum calls for "prioritization of water allocations for environmental flows, conservation of water for future generations, and collaborative decision-making processes" (Baltutis 2012, 1). The National Round Table's report *Moving to Action* cites 3 main priority areas: water data, water value and pricing and collaborative water governance efforts.

2011: National Round Table publishes *Charting a Course: Sustainable Water Use by Canada's Natural Resource Sectors*: 18 recommendations for improving water management and governance.

2010: Canada's Environment Minister releases *Planning for a Sustainable Future: A Federal Sustainable Development Strategy for Canada* containing water quantity and quality targets (Canada Environment 2010d). The National Round Table releases *Changing Currents: Water Sustainability and the Future of Canada's Natural Resource Sectors*. Canadian Council Ministers of the Environment endorses a *Water Action Plan*.

2009: Pollster Nanos Research finds that Canadians believe fresh water is the most important natural resource for Canada's future (Nanos 2009, 1). The Council of Canadian Academies calls for a cooperative government approach to ensure the sustainable development of Canada's groundwater.

2008: The Canadian Water Resources Association shows not what a national water strategy should be, but how to get there (de Loë 2008).

2007: A Conference Board of Canada report recommends actions to improve water management governance. The Gordon Water Group of Concerned Scientists and Citizens publishes their blueprint for change (Gordon Water Group 2007).

Pollution Probe publishes *Towards a National Vision and Strategy for Water Policy in Canada* calling for a greater federal role.

2   Sustainable water management is about making [water] serve present and future needs of humans and other living things…[and] must be ecologically sound, economically and financially feasible, and socially acceptable (Shrubsole and Mitchell (Eds.) 1997, 308).

3   Many disparate Canadian sources contribute knowledge and practical solutions to water problems. There are popular books such as de Villiers' *Water: Our Most Precious Resource* (1999) on problems in many countries, Barlow and Clarke's *Blue Gold* (2003) about the global water crisis and plans of transnational corporations to profit from it, and Barlow's *Blue Covenant* (2007) about the global battle for the right to water. Black's *no-nonsense guide to water* (2004) has a general focus on the water crisis. Black's (2004) *water, life force* is a celebration of water and its roles. Allan Casey's *Lakeland* (2010) is a wonderful example of what I call for in my Chapter 6, as is *Deep Immersion: The Experience of Water* (France (Ed.) 2003). Brooymans' 2011 book *Water in Canada. A Resource in Crisis* presents an alarming picture of the state of our water, including useful hydrological and water quality information. To mention only a few, academic publications include *Practicing Sustainable Water Management: Canadian and International Experiences* (Shrubsole and Mitchell (Eds.) 1997) that discusses policy and practice in Canada, several Commonwealth countries as well as urban perspectives from three U.S. cities; *Resource and Environmental Management in Canada: Addressing Conflict and Uncertainty* (Fourth Edition), brings together contributions from many specialists to focus on management processes, factors that influence decision-making and needed trade-offs (Mitchell (Ed.) 2010); *Eau Canada: The Future of Canada's Water* (Bakker (Ed.) 2007) brings together 26 water experts in 17 essays about Canada's water availability and governance; security; privatization, rights and markets, and future paths; and *Canadian Water Politics* (Sproule-Jones et al. (Eds.) 2008), discusses water use conflicts and the institutional arrangements and reforms needed to meet future challenges. Its focus is the politics of pollution, water withdrawals and in-stream flow protection. Cech's (2004) *Principles of Water Resources* is an American text containing references to Canadian water management.

4   For example, 300 federal positions in the Environment Department are cut with the Spring 2012 budget.

5   Useful sources on climate change include:

Bourque et al. 2007. *From Impacts to Adaptation: Canada in a Changing Climate* <http://adaptation.nrcan.gc.ca/assess/2007/index_e.php>

Canadian Institute of Planners: <http://www.planningforclimatechange.ca/>

Canadian Federation of Municipalities: <http://www.fcm.ca/Documents/reports/PCP/Municipal_Resources_for_Adapting_to_Climate_Change_EN.pdf>

Sandford, R. 2011. *Cold Matters: The State and Fate of Canada's Fresh Water*. Victoria, BC: Rocky Mountain Books

National Roundtable on the Environment and the Economy. 2011. *Paying the Price: The Economic Impacts of Climate Change for Canada*, <http://nrteetrnee.ca/wp-content/uploads/2011/09/paying-the-price.pdf>

UNEP/WMO International Panel on Climate Change <http://www.ipcc.ch>

Chapter 2:

6   The word 'influencers' is used to describe all those who influence water management activities, but who are usually not part of decision-making processes.

7   Water may have been studied longer than any other subject relating to the natural environment (Dorcey 1991, 541).

8   Hydrology is the study of the earth's water, particularly of water on and under the ground before it reaches the ocean or before it evaporates into the air (Mayhew 1997).

9   The supply of fresh water, or Canada's renewable fresh water resources, is represented by water yield. Water yield is the amount of fresh water derived from unregulated flow measurements for a given geographic area over a defined period of time and is an estimate of renewable water (Canada Statistics Canada 2010).

10   For example, *Currents of Change: Final Report / Inquiry on Federal Water Policy* (Canada Environment 1985), and, in date order subsequent: Saunders (1988), Kennet (1991), Booth and Quinn (1995), Pearse and Quinn (1996), Bruce (2006), Saunders and Wenig (2007), Bakker (Ed.) (2007), de Loë (2008) and Sproule-Jones et al (Eds.) (2008)).

11   Water governance is the range of political, social, economic and administrative systems that regulate the development and management of water resources and the provision of water services at different levels of society (Canadian International Development Agency 2003, 6).

12   "Adaptive management" is a way of continually improving management decisions: "When faced with profound uncertainties, take a purposeful step forward, monitor the consequences, learn from the results, and avoid costly failures" (McDaniels 1999, 4).

13   Partners for the Saskatchewan River Basin include representatives from government, First Nations, business, environmental and local interest groups, industry, educational institutions, and individuals in an international watershed including parts of 3 Prairie Provinces and Montana.

14   For example, Zenon Environmental Inc., now acquired by General Electric Company, makes hollow fibre membranes for filtration of deleterious substances in

municipal and industrial water purification treatment plants.

15   Interviews with Jane Henderson in 2003 and 2007.

16   This type of contract gives a private company the exclusive right to operate the water supply infrastructure for an extended period (usually 20 to 30 years) in order that they may recoup their investment.

17   Stakeholder is an individual or group that is affected by or impacts upon an action or intervention (Hickey and Kolthari (2009).

18   Derived from my interpretation of sessions I attended with the late Stephen Haines, *Centre for Strategic Management*.

19   Reflects the work of Kretzmann and McKnight (1993), Cavaye (2000) and others.

Chapter 3:

20   Surface water comes from two main sources: runoff flowing on the ground towards rivers and base flow entering the stream channel from groundwater.

21   A Nation of 17 bands occupying the south central part of BC

22   Following from Roman law, navigable waters of England belonged to the Crown and were available to the public for navigation and fishing. Water not used for these purposes was held by those who owned the banks of streams. These people were said to hold riparian rights.

23   For detailed comparison, see Guelph Water Management Group. 2007. *Characterization of Water Allocation Systems in Canada Technical Report 1*. University of Guelph: Guelph Water Management Group: <http://waterwiki.net/images/a/a8/Technical_Report_1_Oct2007.pdf>

24   The Latin "ripa" means the bank of a lake, river or stream and the etymology of the word "rival" is from one person living on the opposite side of the stream from another.

25   Whether a use is reasonable depends on the size of the stream, the season of the year, the nature of the use and the operations involved.

26   A permit is required for withdrawal of more than 50,000 litres of water (including groundwater) per day unless works were constructed prior to 1961, or the water is for domestic, fire-fighting or farm purposes. These permits are short-term in nature, usually five years.

27   This and subsequent discussion derived from British Columbia Environment, Lands and Parks (1993-9).

28   Terms and conditions include name and location of the stream from which water may be taken or stored, location of the intake on the stream, priority date of the licence, purpose(s) for which the water may be used and the maximum quantity of water to be used or stored, time of the year during which the water may be used, property where water is to be used and to which licence is attached, authorization to construct works to divert and convey the water from the stream to place of use. The licensee must use the water as prescribed in the licence and construct authorized works within the time specified or risk cancellation. Responsibilities include damage resulting from works, even if the license is abandoned, suspended or cancelled; when directed, to keep records of the diversion and use of water and produce them for inspection; water rental fee paid by the specified time or risk cancellation (BC Environment 2006).

29   This discussion of water entitlements is indebted to legal comment: Bartlett 1988; McNeil 1999; Percy 1988; and Kempton 2005.

30   Livia Meret, pers. com.

Chapter 4:

31   Much of this discussion is based on E.C. Pielou's excellent 1998 text *Fresh Water*. Chicago: University of Chicago

32   Subsequent discussion derived from Kohut et al. (1993).

33   Quantity concerns: poor well yields, sometimes insufficient for domestic use; wells flowing without control; pumping of wells lowering the level in other wells; and groundwater mining where the use of wells exceeded their replenishment. Water demand is "concentrated in three subdivisions with lots generally 0.2 hectares in size or less, most of them containing individual ground water wells and on-site sewage systems. ... In 1989, an (Environment) report estimated that the demand for water in these areas had reached as much as 83% of the estimated ground water available. Since then, Hornby's yearround population has increased by at least 19%" (Hornby Island Local Trust Committee 2010).

Quality concerns: improperly abandoned wells (40 reported in 1993); poor well construction (e.g., wells without adequate surface seals allow pollutants (such as from septic fields); deleterious land use practices (e.g., drainage works that affect runoff and paving that reduces recharge); fertilizers that contaminate groundwater; land clearing, logging and road building (soil erosion and disturbance of the natural hydrologic regime); salt water intrusion due to overpumping (groundwater may be contaminated by salt water encroachment, or drawn upward from beneath the fresh water-saltwater interface (upconing)); elevated chloride and hydrogen sulfide levels in dry summer months and elevated coliform counts where subdivision density is high.

34   For information on this study and other work on Hornby Island see the Islands Trust Website.

35   For a more detailed future agenda, see Council of Canadian Academies review of groundwater in Canada (2009).

36   Examples include Waterloo, Ontario's wellhead protection program (FitzGibbon et al. 2009) and the Nova Scotia Ecology Action Centre's Groundswell that is developing community-based groundwater monitoring programs: <http://www.ecologyaction.ca/content/groundswell>.

37   In 2010, BC's Auditor-General found insufficient information about groundwater, lack of protection from quantity and quality problems, insufficient control over access and inadequate authority for local responsibility (BC Auditor-General 2010, 2).

Chapter 5:

38   Ontario and Québec have introduced pricing policies that require water suppliers and waste treatment facilities to charge the full costs of water.

39   These two terms are often interchanged. "Water use efficiency includes any measure that reduces the amount of water used per unit of any given activity, consistent with the maintenance or enhancement of water quality. ... Water conservation refers to efforts made to save water during situations of water shortage, often short term in nature (Tate 2000, 1). "... [E]fforts to increase water efficiency ask how to use water for a given end use; efforts to conserve water ask why we are using water to serve that end use" (Brooks 2006, 527).

40   See <http://www.ec.gc.ca/Publications/default.asp?lang=En&xml=0B6E24B6-0421-4170-9FCF-9A7BC4522C54> for Canadian municipal water rates.

41   The 1992 Dublin International Conference on Water and the Environment was the meeting place for 500 participants, including government-designated experts from a hundred countries and representatives of 80 international, intergovernmental and non-governmental organizations.

42   Various tariff elements include: a connection charge; a fixed charge (flat rate no matter how much water is used); charge by volumetric use, a block charge that provides the same charge for all volumes of water used (often used in agriculture based on area irrigated); a declining block rate (charges less by block as more water is used or a minimum charge); or an increasing block rate (charges more by block as more water is used, or a minimum charge).

43   Xeriscaping is derived from the Greek word "xeros", meaning dry. It originally meant landscaping for dry regions but has been adapted to mean the use of plant materials suitable to an area's climate and practices that minimize water use.

44   Drought is defined as a natural, regular event during which sustained low precipitation and high rates of evaporation cause low water flows in streams and or wells and reservoirs. (BC Land and Water Inc. 2004)

45   "Water rights transfers are a market-based tool that allows all or part of an existing water allocation to be transferred from an existing license holder to another entity for use within the same river basin ... and more specifically within a sub-basin (i.e., along the same river). This type of allocation system does not allow for any *more* water to be withdrawn than under previous allocations; existing rights are simply transferred among users" (Palacios and Brown 2005, 25).

Chapter 6:

46   Instream flow is water flowing within a natural channel although the term may be used to mean the precise quantity and timing necessary to sustain an instream use of water, the "minimum flow" of water or water level necessary to preserve essential values.

47   Turbidity is a measure of the scattering effect that suspended organic and inorganic matter, soluble coloured compounds and microscopic organisms have on light. Water with high turbidity levels has a cloudy appearance and is caused by soil from unstable slopes washing into water.

48   This section inspired by the framework developed for the Auditor-General's report *Protecting Drinking Water Sources* (1999): <http://www.bcauditor.com/pubs/1999/report5/protecting-drinking-watersources>

49   Research has shown a drop in calcium levels in lakes in the Canadian Shield Region due to acid rain and logging (Mittelstaedt 2008, A10).

50   In 2010, fish have returned to near pre-spill levels as a result of work of the Squamish First Nations, three levels of government, volunteer groups and millions of dollars from CN (Pollon 2010, S1).

51   In September 2010, the federal and Manitoba governments signed an agreement to clean-up Lake Winnipeg.

52   Derived from work of Prof. John Taylor, University of Notre Dame and Jeff Alexander's *Pandora's Locks*.

53   Anadromous fish breed in fresh water but live their adult lives in the sea. Examples are salmon, Dolly Varden, lampreys and eulachons.

54   For example, the *Nunavut Waters and Nunavut Surface Rights Tribunal Act*, 2002, defines "instream use" as a "use of waters by a person, other than for a domestic purpose or as described in … the definition "use", to earn income or for subsistence purposes." "Use", in relation to waters, is defined as a direct or indirect use of any kind, including, but not limited to, any use of water power and geothermal resources, any diversion or obstruction of waters, any alteration of the flow of waters, and any alteration of the bed or banks of a river, stream, lake or other body of water, whether or not the body of water is seasonal.

55   British Columbia has used water allocation plans and a water use planning process for existing dams (see Chapter 11). Ontario has an extensive review process relating to 'permits to take water' that must take into account ecosystem needs; Québec has similar authority. British Columbia and New Brunswick have the ability to make sensitive streams unavailable for new licences (BC) or withdrawals (NB). The four maritime provinces have policies to preserve at least 25 percent of the mean annual flow to protect fish habitat (MacDonnell 2009, 1095). Saskatchewan has no specific quantity or quality protections relating to instream flow. Manitoba can suspend or restrict withdrawals to protect aquatic ecosystems while Alberta's Environment Minister can amend a licence to prevent an adverse impact.

56   Results required include maintenance of the stream channel and walls and natural variations in stream temperature, prevention of pollution, conservation of organic debris and forage cycles, protection of fish and wildlife and their habitat and vegetative diversity, and retention of trees and understory vegetation.

57   Retention systems capture runoff and then hold it until it can be safely released. Filtration systems use sand, gravel or other material to capture pollution. Bio-retention systems act like a natural forest ecosystem to treat stormwater runoff (Dzurik 2003 241-249).

58   Value is placed on water because it may be used sometime in the future (option value), because of the pleasure derived from it or because of the ethical duty to preserve it (existence value), because we want water to be available for future generations (bequest value) or because it has the right to exist in the environment (inherent value) (Hannah 1987, 29).

59   In July 2003, "The Water Project" was held in Ontario: 67 galleries and art groups in 30 communities held exhibitions to celebrate and bring public attention to water.

60   Precaution: The precautionary principle denotes a duty to prevent harm, when it is within our power to do so, even when all the evidence is not in. (Canadian Environmental Law Association:

<http://www.cela.ca/collections/pollution/precautionary-principle>

Chapter 7:

61   Where water quality is used in this chapter in relation to drinking water, it means "being free of both disease-causing organisms and chemicals in concentrations that have been shown to cause health problems. Such drinking water has minimal taste and odour, making it aesthetically acceptable to the public for drinking" (Canadian Council Ministers of the Environment 2004, 14).

62   A boil water advisory may be issued (usually by a local public health unit or a water utility) as a precautionary measure if there are emergency repairs to a water line and there is a concern that contamination may occur. Boil water orders are issued if unacceptable levels of disease-causing bacteria, viruses or parasites are found anywhere in the water system from the source to the tap and/or unacceptable turbidity levels are found in the source of the drinking water. Under a boil water advisory or order, water used for drinking, preparing food, beverages, ice cubes, washing fruits and vegetables, or brushing teeth must be brought to a rolling boil for one minute to kill all disease-causing organisms. Water for other purposes such as showering, laundry, bathing, or washing dishes does not require boiling but toddlers and infants should be sponge-bathed with disinfected water (Canada Health 2008).

63   Water Quality Guidelines are established for at least 179 factors including microbiological, chemical and radiological contaminants and physical parameters, such as taste and odour. If drinking water contains greater than recommended concentrations of these substances, then there may be a health risk or the water may be aesthetically unacceptable.

64   For discussion of liquid waste management, see *Wastewater Treatment Made Clear* (Capital Regional District) <http://www.wastewatermadeclear.ca/what/howitworks.htm>

65   For illustration see: <http://www.ccme.ca/assets/pdf/mba_guidance_doc_e.pdf>, p.16

66   For illustration see:
<http://www.health.gov.bc.ca/pho/pdf/phoannual2000.pdf>, p.10.

67   Water is passed through a filter to remove large debris and the sand and grit are settled out. Using coagulation, a chemical like alum is added to the water and when the water is stirred, sticky globs attach to bacteria, silt and other impurities. Then flocculation is used in which the small particles are combined into larger particles. When the water is pumped slowly across a large basin, the floc and solid material settle out in a process called sedimentation or clarification. The water is filtered through layers of granular materials to remove micro-organisms or in more sophisticated systems, through membranes of a polymer skin, and then treated with chemical disinfectants such as chlorine to kill or de-activate organisms present in the water. Chlorination usually involves the addition of chlorine gas or liquid sodium hypochlorite to the water supply as it is both cheap and effective.

68   Political ecology is the study of the relationships between political, economic and social factors with environmental issues and changes (wikipedia.org).

69   See <http://www.bcwwa.org/> for links to other organizations.

70   See Canada-wide *Strategy for the Management of Municipal Wastewater Effluent* endorsed by the Canadian Council Ministers of the Environment in 2009.

Chapter 8:

71   Other references to this case study include Day 1999; Sewell 1969; Smith 1991 and more recent publications of the Fraser Basin Council.

72   The flood profile was created with a program of one-dimensional hydraulic modelling. Downstream of New Westminster, the winter design profile is about 0.3 m higher than the previous profile while upstream of New Westminster the updated profile becomes increasingly higher (Fraser Basin Council 2006b).

73   Schedule AA of the *Official Regional Plan for the Lower Mainland* (1966) delineated the floodplain areas based on local knowledge of the extent of the 1894 flood.

74   Section 187 of the *Municipalities Enabling and Validating Act* since rescinded in favour of reliance on local government bylaws.

75   Floodproofing means maintaining setbacks from water bodies or the toe of dikes and using structural means (which can include the use of fill) to place habitable areas above the Flood Construction Level (the 1 in 200 year flood level plus two feet of freeboard to allow for wave action and other factors).

76   The first $1 per capita is borne by the provincial government, while the next $2 per capita is borne 50% by the government of Canada (the next $2 per capita at 75% and the remainder at 90%).

77   For illustration see:
<http://www.env.gov.nl.ca/env/waterres/flooding/shoal_harbour_pi_map.pdf>

78   Canada Mortgage and Housing Corporation estimates that insurance claims from basement flooding average $140 million a year (Luciw 2011, L3).

79   In 2001 the Canada Office of Critical Infrastructure Protection and Emergency Preparedness (now Public Safety Canada) published an independent expert panel review of managing flood risk (Kumar, A. et al. 2001, 6-9); 16 recommendations for improvement are made, 2 of which are discussed here.

80   Examples in Fraser Basin Council. 2010. *Environmental Protection in Flood Hazard Management*, <http://www.fraserbasin.bc.ca/_Library/Water/report_flood_and_

environmental_protection_2010.pdf>

Chapter 9:

81   Discussion headings inspired by de Villiers' use in his chapter on dams.

82   These figures refer to the number of large dams in each of these countries (World Commission on Dams final report)

83   The International Commission on Large Dams defines dams as "large" if they are more than 15 metres high or 10 metres high with a crest length of more than 500 metres or have a reservoir capacity exceeding 3 million cubic metres (Swiss Re 2003).

84   Illustrations: <http://www.opg.com/power/hydro/howitworks.asp> <http://geoscape.nrcan.gc.ca/whitehorse/yukon_e.php>

85   Richardson (2008, originally 1975) gives more detail.

86   For a detailed accounting, Sean McCutcheon's *Silenced Rivers* should be consulted.

87   For detailed information see Wood, J. 2013. *Home to the Nechako. The River and the Land*. Victoria, BC: Heritage House.

88   See Swift, J. and K. Stewart. 2004. Hydro: The Decline and Fall of Ontario's Electric Empire. Toronto: BTL.

89   Dr. Tom Powers (University of Montana) was appearing before a public forum held in Medicine Hat, Alberta to discuss the proposed Meridian Dam.

90   Here's how a mini-hydro project works on Shook Brook in BC's upper Clearwater River area. A 6 inch water line carries water from the brook about 110 metres to a settling tank where debris is settled out. A 4 inch line then takes the water from the settling tank down 100 vertical metres to a small power house. The water is forced through a nozzle into the wheel of the turbine. The energy produced is carried to a generator and from there to the buildings that are served by the project. This project is connected to the BC Hydro grid. Should the buildings require extra electricity, it is available, or if not, the net energy produced can be transferred to BC Hydro (International Water Power & Dam Construction 2004, 1).

91   This is enough electricity for 1600 homes if operating constantly, but as these plants generate about 80 percent of the time, a back-up source of electricity is required.

Chapter 10:

92 The term 'water export' is used throughout as it is the most easily understood descriptor for 'bulk water removals'. *The Accord for the Prohibition of Bulk Water Removal from Drainage Basins* defines bulk water removal as "the withdrawal and transfer of water out of its basin in quantities which individually or cumulatively could result in damage to the ecological integrity of the system. In general, this could include removals of water by interbasin transfer, pipeline or tanker ship, but not smaller-scale removals such as water packaged in small portable containers."

93 The term "geopolitics" was first used by Kjellen (1901) to encourage change in then current beliefs of European political science that the state was primarily a legal concept. Gradually the term has come to represent the interrelationship between the spatial environment and international politics, particularly the effect of the former on the latter.

94 The human right to water extends to the "right to access to water of sufficient cleanliness and in sufficient quantities to meet individual needs. As a minimum, the quantity must suffice to meet basic human needs in terms of drinking, bathing, cleaning, cooking and sanitation." While the right to water for food or industrial production is not secured, if "a member of a household - most frequently women and girl-children - must walk for hours to fetch daily water, or if fees are so prohibitive that a poor household must sacrifice other essential rights, such as education, health services or food, or else use contaminated water, then individuals of that household are not enjoying their right to adequate water" (World Health Organization 2001).

95 The International Law Association is an association of academics and practising professionals, who study, clarify and develop both public and private international law.

96 The International Law Commission is part of the United Nations and was established in 1948 to develop and codify international law, in accordance with article 13(1)(a) of the Charter of the United Nations.

97 Framework for this section based on Windsor (1992).

98 This project would likely cost $3 trillion (U.S.) today with all environmental approvals (Peterson 2003, 9).

99 In 2009 a reduced proposal suggests harnessing surplus water from seasonal flooding from the Broadback, Waswanipi and Bell Rivers for export purposes (Gingras 2009, 2).

100 The Garrison Diversion, as authorized by the US Congress in 1965, was proposed to irrigate 250,000 acres of arable land in North Dakota, provide water to 14 communities, and enhance recreational opportunities and fish and wildlife. It was to be financed by future power sales from the Garrison Dam. The irrigated

agriculture part of the project was reduced by 50 percent and a principal feeder eliminated, given Canadian opposition and budget constraints (Feldman 1991, 1).

101 "Framing" characterizes how the media emphasize certain aspects of an issue and draw attention to a particular cause or solution to a problem (de Loë 1999; Rogers et al. 1997). Frames determine the limits of "what will be discussed, how it will be discussed and above all, how it will not be discussed" (Altheide cited in Driedger and Eyles 2003, 1281).

102 Six licences had been issued for bulk marine shipment. These licences were issued for terms of 15 years. There were a number of applications to expand these shipments including several suppliers seeking to market water to California. There were a number of licences for bottling mineral and surface water in the Province but no restriction on the size of the container (in one case, containers of up to 5,000 gallons were used). Three licences had been issued for removal of naturally-calved ice from Salmon Glacier near Stewart, BC to be sent to Japanese markets. The Greater Vancouver Water District was supplying water to Point Roberts and the City of Blaine water to Hazlemere in Surrey. There were also a number of groundwater bottling operations that had been removing water for some time. One operation was shipping by tanker truck to Texas for bottling.

103 The *Constitution Act*, 1867 (Sec. 109) grants ownership of all publicly owned "lands, mines, minerals and royalties" to the provinces. Water is not mentioned in these pieces of legislation, but is commonly understood to be part of those resources.

104 Bakenova (2008) traces federal activity in detail from the 1960s.

105 For example, on February 8, 1995, Mr. Bill Gilmour (Comox-Alberni, Reform) moved: "That ... the government should support a policy that Canada's fresh water, ice and snow will be protected so that at all times and in all circumstances Canada's sovereignty over water is preserved and protected". Other similar acts were introduced on a number of occasions since 1997 (Johansen 2001, 10).

106 Illustration of affected area: <http://www.cglg.org/projects/water/docs/GreatLakesCharterAnnex.pdf> Any proposal to divert water (more than 3.8 million litres of water per day, averaged over a 120 day period) from the basin would have to receive unanimous approval of all the signatories to the pact — plus most of the water must be returned to the lakes unpolluted and the amount requested must be reasonable.

107 Such a challenge is the 'Notice of Intent to Submit a Claim to Arbitration' from November 1998 by the U.S. company, Sun Belt Water Inc., but no valid claim has been filed against the Canadian government (Canada Foreign Affairs 2009). In 1991 this company sought a contract to supply water to San Diego in a joint venture with Snowcap Waters of Fanny Bay, BC (who held a bulk export licence). Snowcap held one of 6 licences for marine transport of water, but they contended

that opportunity for expansion was affected by the government's moratorium and subsequent legislation relating to bulk water exports. In 1993 Snowcap, who intended to sell water to Sun Belt, filed a lawsuit for damages against the BC government. This lawsuit was settled out of court in 1996.

108 Series of actions described based on those first mooted by Sewell in 1985 (Windsor 1992, 261).

109 See Canadian Water Issues Council and University of Toronto. Munk Centre for International Studies Program on Water Issues. 2008. *A Model Act for Preserving Canada's Waters.* <http://powi.ca/wp-content/uploads/2012/12/AModel-Act-for-Preserving-Canadas-Waters-2008.pdf>

Chapter 11:

110 For detailed information on these subjects see Dzurik (2003) and Shrubsole (Ed.) (2004).

111 Tools of conflict resolution include negotiation, mediation and arbitration. Negotiation is like a transaction where two or more parties resolve what each shall give and take or perform and receive. Mediation is where a third party attempts to resolve differences using compromise and negotiation. Arbitration is where a third party's decision may be binding. The second technique, mediation, is the most frequently used during planning processes and identified as most useful in resolving rural water dispute situations in southern Ontario (Van Veen et al. 2003, 91, 111).

112 The South Saskatchewan River Basin includes the sub-basins of the Red Deer, Bow and Oldman Rivers and many dams and reservoirs (including the Oldman River Dam) to store water for use in low flow periods, a function of particular importance to irrigated agriculture.

113 Echoing the loss of home to dams are the Lost Villages: 6,500 people lost their homes to Lake St. Lawrence, during creation of the Seaway: <http://activehistory.ca/2012/11/lost-villages-collaboration-and-capturinghistory/>

114 The Columbia River Treaty Committee represented the 260,000 people in the area affected by project construction and included representatives of each of the five regional districts and two members of the Ktunaxa-Kinbasket Tribal Council. Their goal was to have a voice, with the Province of British Columbia, in negotiation and decision making.

115 After completion of plans on 18 critical river systems, the Ontario Waterpower Association negotiated a simpler process (Norris 2003).

116 Watershed: the land that water flows across or through on its way to a common stream, river, or lake

117 See also Joseph 2008, 599

118 France's (2005) *Facilitating Watershed Management: Fostering Awareness and Stewardship* provides many creative ideas for watershed management to meet these goals.

Chapter 12:

119 For illustration see: Natural Resources Canada. 2004. *Climate Change: Impacts and Adaptation*, <http://www.nrcan.gc.ca/sites/www.nrcan.gc.ca.earthsciences/files/pdf/perspective/pdf/report_e.pdf>, p. 63

120 Framework for this discussion from Smith 1999, 272-288.

121 "Negative interdependencies" are uses that impinge on each other; e.g. marinas can destroy fish habitat (Sproule-Jones 2002, 839; 2003, 11).

Appendix:

122 This omission may be partly because the "environment" was not such an important topic in the mid-19th century and partly because the "environment" is such an all-encompassing topic, too broad to give to a single level of government (Irvine 2002, 6).

123 The *Canada Water Act* establishes federal-provincial consultation arrangements and agreements for river basin areas of "significant national interest, permits federal-provincial pollution clean-up programs, regulates nutrients in cleaning products, and deals with administration, public information and advisory bodies." This Act is rarely used today (Bruce 2006, 1).

124 Sewell, W. Lecture notes deposited in the University of Victoria Archives by his family.

INDEX

GEOGRAPHICAL PLACE NAMES
(Except countries, provinces, territories}

Abbotsford, BC, 170
Capital Regional District, BC, 68, 71
Chicago, Ill., 145
Chisabisi, Qué., 132
Cochabamba, 26
Cochrane, Alta., 66-67
Collingwood Harbour, Ont., 90, 169
Columbia River, 128, 143, 150
Devils Lake, 146
Durham, Ont., 71
Edmonton, Alta., 72, 82, 111
Exploits River, 128
Fort George Island, Qué., 132
Fort St. John, BC, 129, 176
Fraser River, 13, 82, 109-117, 150
Gander, Nfld., 174-175, 177
Gander Lake, 174
Gisborne Lake, 153

Great Lakes, 11, 20, 25, 33, 43, 82-85, 111, 143-146, 156-157, 225, 227
Guelph, Ont., 159-160
Halifax, NS, 104, 229
Hornby Island, BC, 51-59, 165, 235-236
Illinois River, 145
Kelowna, BC, 71
Kitchener-Waterloo, Ont., 49, 133
Lake Manitoba, 114
Lake Major, 104
Lake of the Woods, 143, 145
Lake St. Martin, 114
Lake Winnipeg, 84, 146, 237
Liard Basin, 146
Lillooet River, 139
Long Lake, 128
Lower Souris River, 176, 179
Mackenzie River, 33, 43, 81, 146, 225-226

Medicine River, 85
Milk River, 160-161
Mississippi River, 113, 145
Missouri River, 146
Moncton, NB, 26, 27, 98, 104
Montana, 81, 160, 233, 241
Montréal, Qué., 49, 70, 126, 229
Mt. Geoffrey, BC, 52, 54, 56
Nanaimo, BC, 137
Nanaimo Regional District, BC, 120
Nanaimo River, 120
Niagara Falls, Ont., 169
North Battleford, Sask., 40, 98, 100
North Dakota, 146, 242
North Milk River, 160
North Thompson River, 150
Okotoks, Alta., 168-169, 179
Oliver Creek, 82
Pamehac Brook, 128
Peace River, 39, 127,129, 146, 176
Peace River Regional District, BC, 176
Pender Island, BC, 52
Peterborough, Ont., 111, 119
Pockwock Lake, 104
Qu'Appelle River, 174
Rat River, 13
Red River Valley, Man., 82, 113, 14
Richelieu River, 111, 119, 122
Rocky Mountain trench, 145, 146
St. Croix River, 178
St. John River, 82, 170, 174
St. Mary River, 160, 167
Sheyenne River, 146
Shoal Lake, 145
Souris River, 146

Squamish, BC, 82, 98
Starland, Alta., 68
Stephenville, Nfld., 111
Tiny Township, Ont., 28
Thousand Islands, 170
Tofino, BC, 11, 25
Truro, NS, 112
Upper Pitt River, 138
Vancouver, BC, 62, 116
Vancouver Island, BC, 51, 82
Walkerton, Ont., 11, 29, 97-100, 102, 104, 159, 228
Waterton River, 167
Welland River, 169, 179
Whatcom County, Wa., 17
Winnipeg, Man., 49, 68, 109, 113-114, 145-146

GENERAL INDEX

A

aboriginal peoples/First Nations,
　　Action Plan on Clean Water, 227
　　drinking water quality, 97, 100-101, 105, 107
　　fisheries, 85
　　strategies, 183
　　water charges, 63
　　water governance, 229
　　water use, 39
　　water entitlements, 43-46, 185, 189, 235
　　and dams, 125, 131-133, 136, 140, 145, 190; environmental advocacy, 139; floodplain, 190; tar sands, 81; water export, 153; water transfers, 73; water use plans, 171-174, 176-177
　　see also individual First Nations by name
academics, 13-14, 19, 20, 22, 24-25, 92, 107, 178, 183, 242
actors (water), 7, 14-15, 17, 22, 30, 97,114, 163, 182
Adams, Ansel, 9
adaptation strategies, 43, 178, 183, 186, 226, 232, 245
　　Adaptation to Climate Change team, 25

Index   249

advocacy planners, 27, 28

aesthetic values, 18, 21, 35, 64, 80, 91-95, 98, 164, 239

agriculture, 11, 22, 34-35, 39-40, 47, 57, 60, 69-70, 73, 82-83, 94, 112, 115, 117, 129-130, 160, 169, 176, 183, 225, 228, 236, 243-244

Alberta
    Environmental Monitoring Panel, 81
    dams, 136-137
    drinking water quality, 104
    floods, 110
    groundwater supply, 49
    South Saskatchewan River Basin Plan, 167
    tar sands, 80-81
    water allocation, 40, 72-73 (transfers), 238
    WaterSMART, 24
    water supply, 64

allocation (schemes), 12, 14, 18, 20, 33-46, 50, 53, 57, 64, 69, 72-74, 181, 184, 185, 231, 234, 237-238
    apportionment, 42-43
    comparison, 41-42, 44-45
    Northern Authority management, 42, 45
    permitting, 18, 35-37, 41-45, 63, 67, 85-86, 102, 148, 153-154, 161, 234, 238
    prior appropriation, 35, 38-39, 41-42
    Québec civil law, 35, 37
    riparian, 18, 34-44, 72, 79-80, 82, 91, 169, 234
    and aquatic ecosystem, 81, 85- 86, 91, 95, 136; planning, 86, 163, 165, 167

apportionment, see allocation

aquatic ecosystem, 14, 79, 84-91, 94
    damage from major projects, 84
    and drinking water quality, 102
    planning, 165, 168

aquifer, *see* groundwater

Arnstein, Sherry. 28-29

Atwood, Margaret, 11

Auditor-General,
    BC, 98, 236, 237
    Canada, 101, 228

Auditor General, the Commissioner of the Environment and Sustainable Development; the Privacy Commissioner and the Ombudsman, 228

B

Barlow, Maude, 150, 159, 232

Bechtel, 26

Bennett, Premier W.A.C., 128, 150

Berkeley Springs International Water Tasting Competition, 159

best management practices, 69, 90-91, 95, 183

bisphenol A, 159

Blaikie, MP, Bill, 155

"blue baby" syndrome, 170

boil water advisory, 18, 26, 98, 100, 239

bottled water, 15, 101, 142, 147, 158-161, 229
    Bottled Water Association, 159

Bridge River Dam, 139

British Columbia (BC)
    dams, 127-132, 134, 244
    drinking water quality, 104
    drought planning, 69, 70, 237
    flood control, 112-119
    forestry practices, 86
    groundwater, 49, 51-57, 60
    Hydro, 127, 129, 133, 137, 171-173, 179, 241
    Land Commission Act, 112
    surface water allocation, 38-39, 238
    water export, 149-153, 155, 243-244
    water fees, 63
    Water Protection Act, 152
    water use plans, 172-173

BC Provincial Round Table on the Environment and the Economy, 150

BC Water and Wastewater Association, 106

Brooymans, Hanneke, 25, 232

Burstyn, Varda, 158

C

Canada
    Aboriginal Affairs and Northern Development Canada, 101

Action Plans, 227
Agriculture and Agri-Food, 69, 183
Arctic Waters Pollution Prevention Act, 225
Bill C-383, 157
Canada Water Act, 119, 174, 225, 245
Canadian Environmental Assessment Act, 226
Canadian Environmental Protection Act, 84, 226
Constitution Act 1867, 225, 227, 243
Critical Infrastructure Protection and Emergency Preparedness, 240
Department of the Environment Act, 226
Eco-Action projects, 182-183
Economic Action Plan, 27, 186
Environment, 69, 93, 101, 111, 119, 226
federal powers, 84, 225
Fisheries and Oceans, 85, 120
Food and Drug Act, 101, 159
Health Canada, 77, 101
Inquiry on Federal Water Powers, 27, 154, 161, 233
National Hydrology Research Centre, 226
National Pollutant Release Inventory, 102
National Round Table, 45, 231
National Water Research Institute, 226
national programs, 226
national vision for water, 183
Northern Inland Waters Act, 42, 225
On Farm Action Program, 183
RésEau, 227
Sustainable Development Technology Canada, 226
water policy, 27, 50, 154, 182, 226

Canada Mortgage and Housing Corporation, 66, 69

Canada West Foundation, 74

Canadian Broadcasting Corporation (CBC), 25-26

Canadian Council of Ministers of the Environment, 50, 102-103, 155, 228, 240

Canadian Environmental Law Association, 147

Canadian Environmental Solutions, 69

Canadian Heritage Rivers Network, 92

Canadian Water and Waste Association, 24

Canadian Water Network, 22

Canadian Water Resources Association, 24, 150, 231

capacity building, *see* community capacity

Carr, Dawn, 30

Central North America Water Project, 146

Citizens for Water and Power for North America, 150

Clean Annapolis River Project, 90

climate/climate change, 14-15, 18-21, 25, 75, 135, 181-183, 185-186, 226, 232, 233, 237, 245
Kelowna, BC, 71; *and* aquatic ecosystem, 80, 81, 83, 86; drinking water quality, 98, 105, 107; flooding, 114-116, 122; groundwater, 47-48, 59; surface water, 33, 44-46; water export, 144-145, 160

conflict resolution, 244

collaboration, *see* involvement

Colliery Park Dam, 137

Columbia Basin Trust, 171, 179

community capacity, 25, 30, 59, 177, 185, 187, 189

conservation, 12, 14, 23-24, 36, 40-45, 54, 57, 61-62, 64-75, 120, 126, 229, 237, 238;
definition, 236;
and aquatic ecosystem, 88, 90, 92, 94; planning, 163, 167-168, 174, 176; water export, 151-153, 156-158, 161-162

Constable, Guy, 128
consultation, *see* involvement

Council of Canadians, 146, 147, 158

Council of the Federation, 228

Cows and Fish, 91

crisis, 11, 25, 165, 170, 181-2;
and floods, 122; Walkerton, 104; water export, 150, 160

D

Dakota Water Resources Act, 146

Dale, Francis, 150

dams, 13, 15, 17, 19, 38, 125-140, 241, 244
benefits, 127-129
economics, 129-130
failure, 133-134
international guidelines, 138

loss of home, 130-133
problems, 129-136
removal, 137
small hydro, 137-138
tourism and leisure, 134-135;
and aboriginal groups, 139; aquatic ecosystem, 86, 135-136; ethics, 189; flooding, 113-114, 116; planning, 165, 168, 238; water export, 145

Dauncey, Guy, 70

de Loë, Rob, 79

desalination, 7, 73

de Villiers, Marq, 125, 232, 241

disaster planning, 120

Dorcey, Tony, 14

Douglas, Governor James, 39

drinking water quality, 14, 75, 97-107
legislation, 104
multi-barrier approach, 103, 177
non-government, 106
report cards (Ecojustice), 105
and planning, 105-106, 165

drought, 11, 18, 20, 25, 44-46, 62, 69-70, 73, 144-145, 181, 186, 237

Dublin Statement on Water, 63, 182, 236

E

Ecojustice, 24, 97, 105

economic aspects, 14
and dams, 129-130
see also water demand management

education, *see* involvement

efficiency (conservation), 14, 46, 61-75, 94, 106, 157
definition, 236

environmental assessment, 14, 21, 50, 87, 135-136, 153, 183, 226

EPCOR, 72

equity considerations, 67, 94, 144, 189

erosion, 20, 80, 82, 87, 111, 126, 129, 132, 134-135, 170, 174, 179, 229, 235

ethical approach, 12, 15, 29, 34, 44, 46, 64, 91-92, 179, 182, 188-190, 238

European Parliament Water Framework Directive, 184

eutrophication, 135

Evian, 158

export, *see* water export

F

Falkenmark, Malin, 184

Federation of Canadian Municipalities, 229

Fiji Water, 158

First Nations, *see* aboriginal peoples

fish, *see* aquatic ecosystem

flood/floodplain management, 13-15, 17-20, 66, 70, 109-122
events, 110-114
forecasting, 119
human response, 121
insurance, 119-120
integrated management, 117-118
land use policy, 117
structural protection, 116-117
and aquatic ecosystem, 83, 88, 121; climate change, 20; drinking water, 98, 100; pricing, 66

fracking, 82

framing, 147-149, 243

Fraser Basin Council, 118, 168, 240

Fraser River, 13, 150
Flood Control Program, 116, 117

G

Gable, Brian, 156

Garrison Diversion Project, 143, 146, 240

Garrison Reformulation Act, 146

gastrointestinal illness, 99-100

General Agreement on Tariffs and Trade (GATT), 151-152

geography, 12-13, 21, 93, 126, 131-132, 141, 173, 175, 188, 233
historical, 92, 94

252 Index

geopolitics, 141, 242

Gilmour, David, 158

Gordon Foundation, 24

governance, 12, 14-15, 17, 22-23, 30, 174, 178, 182-183, 187, 225-229, 231-232
    aboriginal, 229
    collaborative, 45
    definition, 233
    groundwater, 50-52, 59, 185
    surface water, 35-43
    water, 17, 22
    water allocation, 35
    university programs, 25, 74

GRAND (Great Recycling and Northern Development) Canal, 145

Great Lakes, 11, 20, 25, 33, 43, 82-83, 111, 143-146, 156-157, 225, 227
    Final Report on Protection of the Waters of, 156
    Great Lakes Basin Compact, 85
    Great Lakes Water Quality Agreement, 90 165, 169-170 (plans), 225, 227
    Remedial Action Plans, 90, 169, 226-227

green roofs, 89

groundwater/groundwater management, 12, 14, 17, 19, 21, 26, 30, 36-37, 47-55
    aquifer, 11, 28, 40, 48-50, 54-55, 57-58, 143, 145-146, 151, 155-156, 158, 170, 176-177
    data/information, 49-50, 53-54, 56, 60, 107, 175, 184-185, 226-227, 231
    defined, 47-48
    future directions, 59
    governance, 50
    Guelph, 159-160
    permits, 50, 55
    recharge, 48, 50-52, 54-57, 87-88, 235
    wellhead protection, 55, 236
    wells, 26, 48-51, 53-55, 57, 82, 97, 100, 114, 146, 170, 235, 237
    and Hornby Island, 51-58; Islands Trust, 51

H

Halifax Regional Water Commission, 104-105, 229

harmonization, 158

Henderson, Jane, 26-27, 183

Hornby Island, 51-58, 165, 235-236
    Pilot Project Committee, 52-54, 59

human right to water, 73, 142, 188, 242

hydro-electric power
    see dams, independent power producers

Hydro-Québec, 127, 139

hydrology/hydraulic, 18-21, 85, 141, 174
    definition, 233
    National Hydrology Research Centre, 226

I

independent power producers, 127, 137-138

industry, 11, 22-23, 34, 39, 49, 70, 81-82, 90, 127, 158-161, 187, 233

influencers, *see* actors

infrastructure, 14, 19-20, 23, 100-102, 105, 141-142, 181, 183, 185-186, 189, 234
    Canada Office of Critical Infrastructure Protection, 240
    Instream Flow Council, 79
    and dams, 125-126; floods, 110-111, 116, 121; instream flows, 79-80, 84-86, 237-238; planning, 170-171, 174; water efficiency, 61-63, 65-67, 69, 71-72, 74-75;

integrated resource management, 15, 20-21, 22, 95, 163-165, 167, 169, 179, 181, 184
    see also planning

International Boundary Water Treaty Act of 1909, 143, 146, 156-157, 225

International Commission on Large Dams, 126, 138, 241

International Joint Commission, 27, 134, 142-143, 146, 156-157, 161, 177-178, 225

international legal theories, 143-144

International River Improvements Act, 225

International Water Management Institute, 23

International Year of Freshwater, 187

Internet, 25, 110, 185, 227

intrinsic/intangible values, 11, 13-14, 19, 44, 79, 91-95, 139-140, 181, 188

involvement, 14-15, 27-30, 125, 171, 175, 181
  collaboration, 12, 15, 21-22, 27-30, 45-47, 161, 164, 177, 183, 187, 189, 231
  consultation, 27, 29-30, 37, 153, 172, 167-169, 172, 173, 189
  education, 22, 27, 50, 54, 55-57, 59, 64, 66, 68, 69-74, 91, 94-95, 103, 106, 109, 116, 121, 166, 168, 176-177, 184, 187, 233, 242
  participation, 20, 27-30, 57, 74, 92, 107, 139, 176, 178-179, 187
  see also community capacity, social learning

J

John Day Dam, 150

K

Kean, Ed, 147

Kierans, Thomas, 145

Klahoose, Chief Kathy, 150

Klahoose Nation, 150

Kohut, Al, 48, 235

Kuiper Diversion Scheme, 146

L

Lake Michigan Diversion, 145

Lake Ontario Waterkeepers, 24

landscape aesthetics/design/ecology/preservation, 92, 94, 95, 129, 131-132, 135, 139, 188, 189

Lesage, Premier Jean, 128

licences (water), *see* surface water allocation, 86

liquid waste management, 90, 101-102, 239

local/ regional level, 12, 14-15, 22-23, 26-27, 29, 125, 132, 146, 159-161, 180, 182-186, 188-189, 229, 236
  and aquatic ecosystem, 84, 88-95; drinking water quality, 101-107; groundwater, 47, 50, 52-59; flooding, 109, 114-118, 120-121; planning, 168-169, 172, 176-177; water allocation, 33-35, 44-46; water efficiency 61, 63, 66, 69-72, 75

logging/forestry, 80, 86-87, 129, 133, 171, 175, 235, 237

Lower Souris River Watershed Plan, 176-177, 179

Lucas, Alastair, 35, 46

M

Mackenzie, Alexander, 13

Mackenzie River, 33, 43, 81, 14
  Basin Board and Agreement, 225, 228

Magnum Diversion, 146

Manitoba
  dams, 126-127
  drinking water, 82, 101, 104
  flooding, 113-114
  Garrison Diversion, 146
  groundwater supply, 49
  Hydro, 134-135
  instream flows, 238
  Lake Winnipeg cleanup, 237
  major projects, 84
  water allocation, 39-40
  and water export, 155

Manitoba Wildlands, 146, 160

markets (water), 72, 94, 160, 167, 232

McCurdy Group of Companies, 153

McKnight, John, 30

media/social media, 12-15, 19-21, 24-25, 183, 185-189
  and aquatic ecosystem, 80, 81; dams, 131; drinking water quality, 98; flooding, 109-111, 121; water demand, 69, 71; water export, 141-142, 147-149, 243 (framing)

MHWaters, 147

mining, 49, 80-81, 86, 133

Mitchell, Bruce, 62

Moncton, NB, 26, 27

Greater Moncton Water Treatment Facility, 26

monitoring/enforcement, 45, 50, 54, 74, 81, 85, 87, 89-90, 95, 101, 103-104, 107, 134, 139, 157, 161, 157, 164-166, 170, 173, 175-179, 185, 226, 228, 236

Moosomin Dam, 176

motivation, 26-27, 166, 184

multi-attribute analysis, *see* water value

Multi-National Water and Power, Inc., 150

Munk Centre Program on Water Issues, 25, 244

N

Nestlé, 159-160

New Brunswick
 comprehensive plan, 174
 groundwater, 49
 flooding, 111
 fracking, 82
 instream flows, 238
 planning, 174, 177
 water allocation, 36

Newfoundland and Labrador
 Act to Provide for the Conservation, 153
 flooding, 111
 Hydro, 127
 water allocation, 36
 watershed plan, 174
 and dams, 127-128; water export, 147,153

Niagara Peninsula Conservation Authority, 169

Nisichawayasihk Cree Nation, 139

nitrogen levels, 82-83

Nlaka'pamux First Nation, 140

non-consumptive uses, 73, 92

non-market valuation, *see* water value

North American Free Trade Agreement (NAFTA), 26, 148-152, 154-155, 158

North American Water and Power Alliance (NAWAPA), 145

Northern Authority management, *see* allocation

Northwest Territories
 aquatic ecosystem, 86
 drinking water, 98
 Northern Voices, Northern Waters: ..., 183

Nova Group, 153, 155, 161

Nova Scotia
 aquatic ecosystem, 90
 drinking water quality, 105
 flooding, 111-112
 nitrogen levels, 82
 Nova Scotia Environment Act, 104-105
 water allocation, 36
 and dams, 127

Nunavut
 aboriginal water governance, 229
 instream use, 23
 Water Board, 42

O

Officers of Parliament, 228

Ogallala Aquifer, 146

oil, 11, 51, 80-81, 83-84, 89, 141, 159

official community plans, 55-56, 87, 165, 168

Okanagan Basin/Valley, 69, 105, 174

Okanagan Basin Water Board, 106

Oldman River Dam, 128, 136, 244

Ontario
 agricultural, 94
 aquifers, 49
 bottled water, 158-159
 Children's Water Festivals, 29
 Clean Water Agency, 23-24
 Conservation Authorities (*Conservation Authorities Act*), 90, 168, 174, 229
 drinking water quality, 99-100, 104-106
 infrastructure, 66
 liquid waste management, 102
 photographs, 93
 Power Generation, 127
 Public Utilities Commission, 22
 Royal Ontario Museum, 132
 Society of Artists, 93
 taking water guide, 25

Index 255

Water Resources System, 65
water allocation (riparian system and permits), 35-36, 41, 238
Water Opportunities and Water Conservation Act, 69
water pricing, 63, 67, 236
Water Taking and Transfer Regulation, 153
water use plans, 173, 179
and dams, 127-128, 133-134; media, 22; water export, 153, 157, 159; see *also* Walkerton

Ormsby, Margaret, 13

P

Parr, Joy, 99

Parsons Engineering, 145

participation, *see* involvement

Partners FOR the Saskatchewan River Basin, 22

partnerships, 22-23, 25, 54, 61, 72, 74, 105, 164, 175, 178, 189

Peace Child International, 190

Peace River Regional District, 176

Peace River Watershed Council, 176

Pearse, Peter, 161

Percy, David, 35

permitting, *see* allocation

planning, 12-18, 20-22, 40, 74, 92, 95, 104, 107, 137, 139-140, 156, 175-181, 183-186, 188-189, 226
actors, 21
benefits and challenges, 165
definition, 163-165
disaster, 119
floodplain, 109, 114, 116-118, 121, 168-169
infrastructure, 66
land use, 17, 52, 56, 87-88, 114-118, 182
large scale projects, 170-171
processes, 28, 45
remediation/restoration, 169-170
"soft path", 168
source water protection, 97, 101, 104-107
strategic, 166
surface water allocation, 45-46, 167-168

water demand, 74, 168
water management planning, basis for, 13
water use (dams), 171-173, 238; watershed, 174-176, 188
and aquatic ecosystem, 87-88, 92; groundwater, 50; pricing, 66

POLIS Project on Ecological Governance, 25, 68

political aspects, 11, 17, 22, 25, 29, 59-60, 68, 74-75, 95, 105, 107, 182-184, 189, 240
and dams, 128; flooding, 109, 112, 122; planning, 163, 164, 167, 179-180; water export, 141-142, 147, 157-158

political ecology, 105, 184, 240

pollution, control, 84-85, 89
non-point sources, 80, 94, 174
point sources, 80

Prairie Provinces Water Board, 225, 228

precautionary approach/principle, 149, 188, 238, 239

pricing, 14, 59, 61, 63-72, 74-75, 92, 159-161, 185-186, 226, 231, 236

Prince Edward Island
aquatic ecosystem, 82
groundwater, 36, 49, 50-51
oil tanks, 89

prior appropriation, *see* allocation

private-public partnerships, *see* partnerships

provincial/territorial level, 22, 26-27, 42, 47, 74, 127, 136, 166-167, 182, 225-228, 240, 245
strategies, 183
and aquatic ecosystem, 79, 84-86, 92, 95; drinking water quality, 100-106; flooding, 116-119, 122; groundwater, 50, 53, 55, 57-59; planning, 169, 171, 174; water allocation, 39-40, 43-44, 184; water efficiency, 63, 66, 68-71, 74, 186; water export, 149-158, 160

public health, 13, 19-21, 79, 100, 104, 165, 239

public participation, *see* involvement

Purdy, Alan, 93-94

Q

Québec
 flooding, 111, 122
 Hydro-Québec, 127, 139
 instream flow, 238
 nitrogen levels, 82
 St. Lawrence Vision 2000, 90
 water allocation, 37, 44, 185
 water quality guidelines, 102
 and dams, 127-134; pricing, 236; water export, 155, 157-158
Queen's University Water Initiative, 23

R

rainwater management, 54, 70-71, 83, 89
Real Estate Foundation of BC, 176
Revelstoke Dam, 28, 126, 129
Rideau Canal, 128
riparian, *see* allocation
Roué, Marie, 139
Royal Bank of Canada Blue Water Project, 22
Royal Society of Canada, 81

S

Saik'uz and Stellate'en First Nation, 136
St. Lawrence Seaway, 128, 144, 170-171, 179, 226, 244
St. Lawrence Vision 2000, 90
Salteau First Nations, 176
Sarton, May, 47
Saskatchewan
 instream flows, 238
 drinking water quality, 100, 105
 water allocation, 40
 Watershed Authority, 176
 and dams, 127; water export,155
Savoie, Donald, 24
Science Council of Canada, 69

Schindler, David, 64, 81
Sechelt Indian Band (shíshálh), 138
sewer use bylaws, 89-90
Sierra Legal Defence Fund *see* Ecojustice
social impact assessment, 131, 139-142
social learning, 47, 58-59, 177, 187, 189
social values/capital/security, 11, 12, 15, 17, 23, 29-30, 46, 59, 64-67, 90, 125, 130, 132, 135, 162-167, 169, 179, 181-184, 186-187, 189, 229, 240
South Saskatchewan River Basin Plan, 167
spills, chemical, 81-82, 89, 237
Squamish First Nation, 237
stakeholders, 12, 21, 27, 40, 166, 184
 and groundwater, 53; planning, 166, 169, 172-173, 175
State of the Environment reports, 102
St'àt'imc First Nation, 139
Stenson, Fred, 61
Stieglitz, Alfred, 9
Stockholm Water Institute, 184
stormwater (management), 50, 83, 87, 89, 95, 102, 176, 238
strategic approaches/planning, 15, 17, 55-57, 66, 74, 86, 104, 118, 122, 139, 155-156, 163-164, 166-170, 174-179, 181-183, 185-186, 228, 231-232, 234
Stream of Dreams, 91
subdivision approval, 87-89, 113, 117-118, 235
surface water, 14, 21, 234
 allocation, *see* water allocation schemes
 availability, 19, 33, 50, 59, 81
 conflicts, 34
 data/information, 36, 74, 95, 107, 144, 156, 173, 175, 184-185, 226, 231
 sustainability, 12, 14-15, 30, 181-182
 treatment, 97
 transjurisdictional nature, 225
 uses, 34-35
 and aquatic ecosystem, 80-82, 87; economic good, 64; efficiency, 63, 70; ethics, 188;

groundwater 49-50, 94-95, 159; local government, 94, 183; planning, 163, 179-180; private sector, 187; water export, 141-144, 156, 243

sustainability, 14-15, 30, 33-34, 46, 58-59, 64, 70, 141, 144, 163, 179-183, 187-188, 231
definition, 12, 232
checklists, 94
and planning, 58

T

think tanks, 25

Thomas, Michael, 121

Thomas Peacock, Anna 147-149

Tla'Amin First Nation, 137

transboundary issues, 43, 52, 143, 156, 178, 227-228

transfers (water), 72-74, 88, 154-155, 167, 185
definition, 237

tritium, 98-99

turbidity, 80, 85, 98, 103, 135-136, 237, 239

U

United Nations Environment Programme, 11, 138

United Nations General Assembly, 143, 247

United Nations Water Program, 11

US Army Corps of Engineers, 170

V

Vancouver, BC, 62, 116
Greater Vancouver Water District, 98, 243

Veolia North America, 26

volunteer activities, 12, 14, 19-22, 24-25, 53, 55, 58-59, 85, 90-91, 95, 110, 119, 183, 187, 189, 237

Vyvyan, Clara, 13

W

Waddell, Ian, 158

Walkem, Ardith, 140

Walkerton, 11, 22, 27, 97-100, 102, 104, 159, 228
Clean Water Centre, 106
political aspects, 22
Ontario artists, 93

Walter and Duncan Gordon Foundation, *see* Gordon Foundation

water availability, 14, 17, 19, 27, 33, 36, 54, 95, 161, 165, 167, 181, 232

water conflict, *see* water export

Water Conservation Trust of Canada, 85

water demand/water demand management, 14, 17, 44-45, 64, 72, 74-75, 162, 181, 186
and pricing, 67-68; planning, 163, 165, 167- 168, 174

water ethic, *see* ethical approach

water export, 10, 12, 15, 18, 19, 27, 139-162, 165, 181, 226, 244
bottled water, 158-160
concerns, 142
conflict, 141-142
definition, 242
environmental issues, 144
federal involvement, 154-158
legal theories, 143
past proposals, 145-147
provincial involvement, 149-154
and media, 147-149

Water Governance Program, 25, 74

Water Institute, 25

water management
actors/influencers, 18, 21-27, 182, 189, 233
changes in, 109
costs, 65-66
challenges of, 163
constituencies, 21
effectiveness/success, 5, 71, 182
ethical approach, 92
framework, 18-20
history, 13, 18
involvement in, 27-30, 186
issues, 18
meaning, 17, 58, 141

partnerships, 22
planning, 13, 163-180
politics of, 47
sustainability, 58
see *also* groundwater

Water Quality Guidelines and Index, 102, 104, 239

water value, 18, 29, 46, 67
    aesthetic, *see* aesthetic values
    artistic, 93-94
    ecological/environmental, 85-86, 89, 94
    economic, 19, 62-65, 69, 72
    hedonic, 92
    inferential, 92
    intrinsic/intangible, 11, 13-14, 18, 19, 79-80, 91-95
    multi-attribute analysis, 93
    recreational, 44
    social, 17, 67

Waterkeepers Alliance, 91

Welland River Restoration Committee, 169-170

Wellington Water Watchers, 159

wells, *see* groundwater

West Moberly First Nations, 176

Western Canada Water, 166

Western Water Augmentation Scheme, 146

Wolf, Aaron, 161

World Commission on Dams, 132, 138,

Wrobel, Janusz, 93

XYZ

xeriscaping, 69-70, 237

Youngerman, Connor, 190

Yukon
    aboriginal water governance, 229
    climate change, 20
    Mackenzie River Basin Board, 228
    Water Board, 42

zebra mussels, 83

Zenon, 73, 233

zoning, 50, 88, 89, 117, 122, 164-168

GREEN FRIGATE BOOKS

"THERE IS NO FRIGATE LIKE A BOOK"

Words on the page have the power to transport us, and in the process, transform us. Such journeys can be far reaching, traversing the landscapes of the external world and that within, as well as the timescapes of the past, present and future.

Green Frigate Books is a small publishing house offering a vehicle—a ship—for those seeking to conceptually sail and explore the horizons of the natural and built environments, and the relations of humans within them. Our goal is to reach an educated lay readership by producing works that fall in the cracks between those offered by traditional academic and popular presses.